W9-AHT-699

The Science
in Science Fiction

ALSO BY ROBERT W. BLY

The Unauthorized Star Trek Quiz Book

Why You Should Never Beam Down in a Red Shirt

Comic Book Heroes: 1,001 Trivia Questions
About America's Favorite Superheroes

The "I Hate Kathie Lee Gifford" Book

The Elements of Technical Writing

83 SF Predictions That Became Scientific Reality

THE
Science IN
Science Fiction

Robert W. Bly

CONSULTING EDITOR
James Gunn

BENBELLA BOOKS, INC.
Dallas, Texas

BenBella Books, Inc.
6440 N. Central Expressway, Suite 617
Dallas, TX 75206
www.benbellabooks.com
Send feedback to feedback@benbellabooks.com

PUBLISHER: Glenn Yeffeth
EDITOR: Shanna Caughey
ASSOCIATE EDITOR: Leah Wilson
DIRECTOR OF MARKETING/PR: Laura Watkins

Printed in the United States of America
10 9 8 7 6 5 4 3 2 1

Library of Congress Cataloging-in-Publication Data

Bly, Robert W.
    The science in science fiction : 83 SF predictions that became scientific reality / by Robert W. Bly.—Benbella Books ed.
       p. cm.
    ISBN 1-932100-48-2
    1. Science—Forecasting—Popular works. 2. Science in literature. I. Title.

Q162.B65 2005
501'.12—dc22

                                                              2004024663

Text design and composition by John Reinhardt Book Design
Cover design by Todd Bushman
Cover art by J. P. Targete

Distributed by Independent Publishers Group
To order call (800) 888-4741
www.ipgbook.com

For special sales contact Laura Watkins at laura@benbellabooks.com

"All good ideas have been thought of by somebody before they are realized."

 —Arthur C. Clarke, *Tales from the White Hart* (1957)

"Science is a project in a constant state of revision. Theories are tweaked, probabilities adjusted, limits pushed, elements added, maps drawn. And every once in a while, a whole chapter gets a rewrite."

 —*Scientific American*, March 2003, p. 10

"It is less than five hundred years since an entire half of the world was discovered. It is less than two hundred years since the discovery of the last continent. The sciences of chemistry and physics go back scarcely one century. The science of aviation goes back forty years. The science of atomics is being born. And yet we think we know a lot. We know little or nothing. Some of the most startling things are unknown to us. When they are discovered, they may shock us to the bone."

 —Donald A. Wollheim, "Mimic" (1942)

"Science fiction writers have been making predictions for generations now, and because the accuracy of the forecast is only as good as the quality of the information being used, the predictions of science fiction writers have generally been better than those of anyone else's—including the complex computerized 'world models' of the scientists who call themselves futurists."

 —Ben Bova in *The Book of Scientific Anecdotes*,
  Adrian Berry, ed. (1993)

"I think science fiction is a very important kind of literature...and I've struggled back and forth between my desire to make science fiction into a visionary literature of great emotional and literary intensity, and the publisher's desire to make a lot of money by selling a popular entertainment."

 —Robert Silverberg, *Publishers Weekly*, May 3, 2004, p. 176

"All great discoveries had, at first, a devastating effect on the state of the world and on its image in our minds. They shattered it and introduced new conditions. They forced the world to move forward. But this was possible only because the discoverers were not afraid of the consequences of their discoveries, no matter how terrifying these were to all those who wanted to preserve the world as it was and hang a big notice on it, saying: Please do not disturb."

   —Edward Teller, *In the Matter of J. Robert Oppenheimer* by Heinar Kipphardt (1968)

"Scientists do not, in the ordinary course of their investigations, consider how their quest for knowledge will affect the future and human existence. Science fiction was the literary response to the fact of man-made change and the possibilities of catastrophic or fundamental changes in the Earth, the Solar System, the galaxy, or the universe. Mostly it came to concern itself with the effects of scientific and technological changes on people in the real world, and thus began to act not so much as the champion or critic of science but as the analyst of scientific possibilities and their human consequences."

   —James Gunn, *Welcoming Remarks* at the 2004 Campbell Conference

"We're living in a science fiction world. The headlines of our newspapers trumpet cloning experiments, the teleportation of atoms, computers controlled by brainwaves, robots that walk and play musical instruments, private planes that carry passengers to the edge of space, space probes that rendezvous with asteroids, and implantable devices that restore hearing to the deaf. When miracles like these are the baseline, how can fantasy measure up?"

   —Eric S. Rabkin, *BusinessWeek*, October 11, 2004, p. 206

For Heidi, Humphrey, and Brownie—in loving memory

Contents

The Greatest Science Ideas That Originated in Science Fiction

Acknowledgments

THANKS TO MY EDITOR, Glenn Yeffeth, for making this book much stronger than it was when it first crossed his desk, and his extraordinary patience waiting for it to get there. And thanks to my agent, Bob Diforio, for getting the book to Glenn in the first place. Special thanks to science fiction writer and historian James Gunn for his helpful review of the manuscript and for pointing me to many sources I had forgotten or was unaware of.

I acknowledge the numerous books, short stories, articles, movies, and TV shows that served as reference materials for this book, but in particular I must single out the most comprehensive reference work on science fiction, *The Science Fiction Encyclopedia* by John Clute and Peter Nicholls (St. Martin's Griffin, 1995 edition).

I also want to thank my two research assistants, Amy Fairchild and Nurit Mittlefehldt, for getting me what I needed promptly and efficiently, and finding some gems I would not have uncovered on my own.

Introduction: Science Fiction as Science Prediction

T RUTH IS STRANGER THAN FICTION. Or so the saying goes.
But in science fiction, it often takes truth a long time to catch up
with fiction; many of the most fascinating ideas in science originated not in the laboratory, but from the minds of imaginative science
fiction writers.

As a twelve-year-old boy, Simon Lake read Jules Verne's *Twenty Thousand Leagues Under the Sea* (1869), which inspired him to become a marine engineer. In 1895, Lake designed and built the Argonaut, the first
submarine to operate successfully in the open sea. Jules Verne sent a
cable to Lake congratulating him on transforming "a work of my imagination" into a reality.

In fact, submarines had been speculated about more than two hundred years before Verne's novel, by John Wilkins in *Mathematical Magic*
(1648). David Bushnell had built an experimental submarine in 1775
and Robert Fulton in 1800, but it was not until Lake that submarines
became practical.

Arthur C. Clarke, author of *2001: A Space Odyssey,* invented communications satellites in an article in 1945. Robert Heinlein practically did
the same with nuclear warfare. A large number of the scientists responsible for sending rockets to Mars say their inspiration came from Ray
Bradbury's *The Martian Chronicles,* and astronomer Carl Sagan credited
science fiction for sparking his interest in science.

Aldous Huxley and many other writers predicted cloning decades
before Dolly the sheep. As sf historian and writer James Gunn observes:
"Many inventions, from Buck Rogers' backpack rocket to robots, lasers,
and computers, have first been described in science fiction stories. But
the literature owes an equal debt to science, from which it drew not
only inspiration but many of its ideas." Though most predictions made
in science fiction don't come true, many of those that *do* come true have
been described with uncanny accuracy, decades before the science is
proven to be correct or the technology feasible.

For instance, Robert Silverberg, in his 1976 novel *Shadrach in the Furnace*, describes an artificial liver in great detail:

> Since Genghis Mao will be liverless for four to six hours, an artificial liver must be used to sustain him during the operation. But no wholly artificial liver has ever been perfected, not even now, after more than fifty years of organ-transplant technology.
>
> The squat cubical device Warhaftig employs is a mechano-organic composite: pipes, tubes, pumps, and electrodialytic filters keep the patient's blood properly pure, but the basic biochemical functions of the liver, having thus far proven impossible to duplicate mechanically, are performed by the naked liver of a dog, resting in a bath of warm fluid at the core of the operation.

Notice that the artificial liver is an external device, and the liver functionality is provided by a donor organ—specifically, a dog's liver kept within the machine's chamber.

As I write this in 2005, I have on my desk a report from a company called HepaLife describing the artificial liver it is developing. The parallels between this technology and Silverberg's description are incredible:

> The patient's blood is routed into a vessel containing HepaLife's patented line of "PICM-19" swine liver cells. The cells detoxify and purify the patient's blood, doing the job his liver cannot. The cleansed, treated blood is then returned to the patient's body.
>
> Within the vessel, two plates, comprised of 12-inch square glass or plastic sheets, are sealed at the edges and spaced about 200 micrometers apart. Mature PICM-19 liver cells grow on the inner plate surfaces.
>
> The artificial liver device has an intake and outflow manifold, allowing for the flow of blood or plasma to pass through the plates. The patient's blood flows between the plates, where it interacts with the PICM-19 cells.
>
> The cell culture plate is constructed with a semi-permeable membrane so fluid such as media, blood, or plasma may be effectively circulated through the apparatus for liver dialysis. The membrane contains the PICM-19 within the vessel so they do not escape into the patient's circulatory system.
>
> A single plate can contain approximately one billion adult PICM-19 cells, all alive and performing the many functions of a healthy liver.

The major difference is that while Silverberg uses the whole liver of a dog, HepaLife's artificial liver uses a colony of liver cells from a pig. Notice how Silverberg correctly predicted that, even in a bionic liver, the liver functionality would have to be performed by living cells, and could not be duplicated mechanically—which is the line of research virtually all scientists in this field are pursuing today. It amazes me that Silverberg took the time to think this through *and* got it right, despite the fact that he is neither an M.D. nor (as Isaac Asimov was) a biochemist.

Perhaps Silverberg should get a royalty when the artificial liver is patented. According to James Gunn, sf editor Hugo Gernsback once campaigned to allow science fiction writers' ideas to be patented, on the basis that getting the idea is ninety percent of the invention. (As an engineer, I'm not sure I agree.)

Here's another example of a science fiction writer getting the idea first: in his 1953 short story "The Golden Apples of the Sun," Ray Bradbury describes a space mission in which a rocket is sent to collect a small chunk of material from the Sun using a solar "scoop."

Almost half a century later, NASA launched the *Genesis* probe into deep space, where it spent twenty-six months collecting matter from the Sun before returning to Earth (the probe crash-landed in a Utah desert when its parachute failed to open, but some of the material was recovered successfully). The solar material consists of ions—atoms stripped of many of their electrons—that are emitted by the Sun in what is known as the "solar wind." The *Genesis* mission's purpose: to increase our knowledge of solar chemistry.

The Science in Science Fiction is written to be a layperson's guide to science ideas, inventions, important events, and social changes that appeared in science fiction before they were eventually embraced or conceived of by the scientific community—from bionics and black holes to warp factors and wormholes. Some of these scientific ideas, such as space stations and cloning, have become almost commonplace—they are that well established. Others, such as antigravity devices, may never come to pass, but the possibility of their creation has become somewhat more probable due to recent developments in scientific theory or new laboratory discoveries.

For a science fiction concept to make our list of "great science ideas that originated in science fiction," the notion had to fit one of the following scenarios:

1. The idea clearly originated in science fiction, then was later discovered or achieved by scientists (e.g., putting a man on the Moon).

2. The idea originated in science, but was brought to the public's attention by science fiction writers (e.g, black holes).
3. The idea originated in science fiction, and we are getting increasingly close to making it a reality (e.g., flying cars, jet packs).
4. The idea originated in science fiction and has not yet been realized in actuality by science, but is at least theoretically possible according to current scientific findings (e.g., time travel, teleportation).

This book takes a look at dozens of the most important science ideas found in science fiction. Each chapter presents a different science idea that originated in science fiction and includes (depending on the availability of accurate information):

- The author(s) who came up with the notion.
- The book, story, or movie in which the idea was introduced.
- How the idea evolved from an sf story into a plausible scientific fact or theory.
- The science behind the idea (what it is, how it works).
- The research that has been done to support or prove the idea.
- Future developments (e.g., will matter transport, as portrayed in the transporter of *Star Trek* and the "jaunt" of Alfred Bester, ever be a reality? Answer: the likelihood is looking better and better; see my chapter on "Teleportation" for the full story).

Of course, one could write an entire book, or several books, about many of the scientific ideas I cover in my chapters (many people in fact have done just that). So if you'd like to explore any topic in greater depth, you can turn to the Bibliography at the back of the book and check out the books, articles, and Web sites I used as my source materials. I also recommend you spend some time at www.technovelgy.com, the premiere Web site on inventions and ideas originating in science fiction books and movies.

However, I do have one favor to ask: as you read, you may think of additional science fact/science fiction ideas that could have been included in this book, but that I omitted. So if you know of an important scientific discovery that first originated in science fiction, why not send it to me so I can share it with readers of the next edition? You will receive full credit, of course.

You can reach me at:

Bob Bly, 22 E. Quackenbush Avenue, Dumont, NJ 07628
rwbly@bly.com

Alternate Energy

LTERNATE ENERGY SOURCES are a perennial science fiction favorite.

Perhaps the most famous novel written on this theme is Isaac Asimov's *The Gods Themselves* (1972), in which scientists generate energy to provide power for the utility grid in both our universe and a parallel universe (called the para-universe) by exchanging matter between the two universes. Here's how Asimov describes the technology:

> The strong nuclear interaction, which is the strongest known force in our universe, is even stronger in the para-universe; perhaps a hundred times stronger.
>
> This means that protons are more easily held together against their own electrostatic attraction and that a nucleus requires fewer neutrons to produce stability.
>
> Plutonium-186, stable in their universe, contains far too many protons, or too few neutrons, to be stable in ours, with its less effective nuclear interaction.
>
> The plutonium-186, once in our universe, begins to radiate positrons, releasing energy as it does so, and with each positron emitted, a proton within a nucleus is converted to a neutron.
>
> Eventually, twenty protons per nucleus have been converted to neutrons and plutonium-186 has become tungsten-186, which is stable by the laws of our own universe.
>
> In the process, twenty positrons per nucleus have been eliminated. These meet, combine with, and annihilate twenty electrons, releasing further energy, so that for every plutonium-186 nucleus sent to us, our universe ends up with twenty fewer electrons.

So far, it doesn't look like we're going to solve the global energy shortage with the help of extraterrestrials, let alone parallel universes. But what about alternate energy sources—wind, water, solar, and geothermal—right here on Earth?

Roger Conrad, an energy analyst and editor of *Utility Forecaster*, a utility industry newsletter, predicts that by 2010, one-third of our energy will be generated by burning coal, half will come from natural gas, and one-tenth from nuclear energy. Only one percent of power on the utility grid will be produced by burning oil, and five percent from hydroelectric power and other "renewable" energy sources such as wind and geothermal.

Windmills have been around for more than 1,300 years, used first in Persia in 700 A.D. and introduced in Europe by France in 1180 A.D. The problem with windmills, as with other alternate energy sources, is the dependence they create on the whims of the weather: windmills don't generate power when the air is still. On the other hand, wind is a renewable energy resource. Wind is produced by uneven heating of Earth's atmosphere by the Sun, so as long as the Sun shines and Earth has an atmosphere, there will be wind.

Waterwheels, an even older source of alternate energy, don't depend on the weather for operation, but are bound to locations near rapidly flowing water. Luckily, there is one place on Earth where water is always flowing with an incalculable amount of kinetic energy: the ocean. If you've ever gone swimming in the ocean and been pounded by waves or dragged out to sea by the undertow, you've felt this force. Now engineers at several companies around the world are building devices that can convert the force of ocean waves, swells, and tides into usable energy.

One such device, the "Seadog" wave pump, uses the mechanical energy of ocean swells to pump seawater to land-based hydroelectric turbines. The water moves through the turbines, spinning the blades to generate electricity, and then is returned to the sea. But the turbines still have to be near the shore; you can't run a Seadog wave pump in Arizona. Another device meant to generate electricity from ocean waves consists of buoys tied to generators. As waves pass, the buoys bob up and down, creating a flow of hydraulic fluid. The fluid flow powers an electrical generator sealed in a water-proof housing on the ocean floor. A cable carries the electricity from the generator to a transformer on the shore. The buoys are fifteen feet in diameter and forty feet long. Each can generate up to fifty kilowatts of electricity, enough to light up to fifty homes. Naturally, the areas most likely to benefit from this technology are towns and cities on or near the ocean.

In 2003, a wave-based power-generation system began providing electricity to homes on the Arctic tip of Norway. Tidal currents in the Kvalsund sea channel turn the thirty-three-foot-long blades of a turbine bolted to the sea floor. During the twelve hours each day when the tides rise and fall, they send water in and out of the channel at a speed of eight feet per second. The Kvalsund turbine system cost about $11 million. It generates approximately seven hundred thousand kilowatt hours of energy a year, enough to light and heat about thirty homes.

Arthur C. Clarke predicted the generation of electricity from the ocean in his 1962 short story "The Shining Ones." He envisioned thermal energy, instead of kinetic, being converted to electricity:

> The most ambitious attempt yet made to harness the thermal energy of the sea... [depends] on [a] surprising fact: even in the tropics the sea a mile down is almost at freezing point. Where billions of tons of water are concerned, this temperature difference represents a colossal amount of energy—and a fine challenge to the engineers of power-starved countries.
>
> For over a hundred years it had been known that electric currents flow in many materials if one end is heated and the other cooled, and ever since the 1940's Russian scientists had been working to put this "thermo-electric" effect to practical use.
>
> Their earliest device had not been very effective—though good enough to power thousands of radios by the heat of kerosene lamps. But in 1974 they had made a big, and still-secret, break-through. Though I fixed the power elements at the cold end of the system, I never really saw them; they were completely hidden in anticorrosive paint. All I know is that they formed a big grid, like lots of old-fashioned steam radiators bolted together.

The problem with thermal energy conversion is that it can only work in areas where there is sufficient heat in the ground or water that can be efficiently converted to electric power. Unfortunately, only a few areas of the world have sufficient geothermal activity to make geothermal energy a significant source of alternate energy.

Australia has large reserves of geothermal energy in granite buried within six to ten miles of the surface. One cubic kilometer of hot granite at 250 degrees Centigrade holds geothermal energy equivalent to forty million barrels of oil. Iceland also has rich geothermal fields as well as many fast-flowing rivers ideal for hydroelectric power generation. As a result, two-thirds of Iceland's energy is produced from alternate (renewable) sources, mainly water and geothermal—a feat unmatched by any other nation.

Wind and other renewable energy sources generate one-fifth of Denmark's electricity. And researchers in Thailand are developing a method to convert palm oil into a replacement for diesel fuel. New power plants are designed to burn wood chips, cornstalks, and husks instead of oil, coal, or natural gas.

In the movie *The Matrix*, a band of humans fighting for freedom against the computers that have taken over the Earth live in Zion, an underground city, located close enough to the Earth's core that it can draw its energy from the heat of molten rock flowing nearby. The geothermal energy is sufficient to run a city with a population of approximately 250,000 people.

Solar energy is also a popular topic, among both sf writers and those seeking renewable energy sources. James Gunn describes a solar energy project in "Child of the Sun" (1977):

> "Everybody knows about the Solar Power Project," Johnson said. "It's no secret."
>
> "I guess not," the engineer admitted. He looked at the metal table with its printed wood grain as if he wished it were a drawing board. "This is an experimental project, and we've demonstrated that we can get significant amounts of power out of solar energy."
>
> "Then why is the project still experimental?" Johnson asked.
>
> The young man at last found something to do with one hand. "Well," he said, rubbing his chin and making the day's stubble rasp under his fingers, "there's one problem we haven't solved."
>
> "The daylight problem?"
>
> "No. Energy can always be stored by pumping water, electrolyzing it into hydrogen and oxygen, with batteries or flywheels. The problem is economics: it's cheaper to burn coal, even if you toss in the cost of environmental controls and damage. Almost one-fourth as cheap. And nuclear power costs less than that. Other forms of solar power, including power cells for direct conversion of sunlight into electricity, are either less efficient or more expensive."

Even though the Earth is ninety-two million miles from the Sun, we receive about eighty-five trillion kilowatts of constant energy from the Sun—equivalent to burning 1,150 billion tons of coal per year. While solar energy depends on sunny days, there are many areas of the world that get enough sunny weather to make solar power a viable alternative to—or, more commonly, a supplement for—utility grid power.

The most effective solar power systems use some kind of dish or reflector to capture and concentrate the Sun's rays. In a "parabolic-trough" system, a series of long, rectangular, curved mirrors are tilted toward the Sun. The mirrors focus the sunlight on a pipe that runs down the center of the trough, heating oil flowing through the pipe. The heat from the oil is transferred to water, boiling the water to produce steam. The steam is used to generate electricity. The cooled oil is returned to the pipe so it can be reheated again by the solar rays.

Another technology, the *solar cell*, converts the Sun's rays directly into electricity through the photovoltaic method. The first solar cell, introduced by Bell Telephone Laboratories in 1954, was made from a small silicon wafer.

In a solar cell, photons strike the silicon, causing electrons in the material to move. The movement results in uneven distribution of electrons in the silicon. When a wire is connected between the two sides of the cell, electrons flow from where the concentration is high (the negative pole) to where there are fewer electrons (the positive pole), creating an electric current. The photovoltaic process does not deplete the silicon, so the solar cell can generate current indefinitely without wearing out. Even though silicon is one of the most plentiful elements, photovoltaic cells are expensive to manufacture. This is because they require an ultrapure silicon that must be refined through extensive processing.

One of the reasons for the limited commercial application of solar energy is the cost: inefficiencies in the solar conversion process make solar energy, on a dollars per kilowatt basis, much more expensive than coal, natural gas, oil, or nuclear. But the rising cost of conventional energy, combined with newer, more efficient solar conversion technologies, is rapidly making solar energy a cost-competitive alternative to fossil and atomic fuels.

Solar cells are also shrinking and becoming lighter in weight, enabling the manufacture of portable solar energy converters. A European Union research team has developed light, flexible solar panels that are not much thicker than photographic film and can easily be applied to everyday fabrics. They can also be fabricated in rolls that can be cut and affixed over a roof as a solar energy source.

Burning garbage—biomass in the politically correct term—is a feasible if problematic source of energy. One of the problems is efficiency:

burning the biomass in such a way that a large portion of the energy is transformed into usable power. A second is pollution: not releasing or creating pollutants as the biomass is burned.

"Gasification" technology is one method of converting biomass into energy. Gasification systems produce energy by burning carbon-based waste products such as bio-waste, agricultural wastes, and municipal sewage. The organic material in the waste reacts with steam and oxygen at high temperature and pressure, which chemically converts it to syngas. The biomass is incinerated in such a way that burning it produces "synthesis gas," or *syngas* for short. Syngas contains hydrogen, carbon monoxide, carbon dioxide, and nitrogen. It has a high energy content and can be compressed and stored for later use, such as driving a turbine engine. Or, it can be catalytically converted to produce ethanol, natural gas, or anhydrous ammonia. The extreme temperatures incinerate the nonorganic materials in the garbage, such as metals and plastics, into ash. The ash is inert and has a variety of uses in the construction and building industries.

Wastewater can also be converted into energy. In this process, the water is piped into a fuel cell with graphite electrodes and a catalyst membrane made of carbon, plastic, and platinum. Microbes within the wastewater generate free electrons as their enzymes break down sugars, proteins, and fats. During the process, up to seventy-eight percent of the waste products in the water are removed. The free electrons, meanwhile, produce ten to fifty milliwatts of power per square meter of electrode surface.

Blue Spruce Farm, a dairy farm in Bridgeport, Vermont, uses a biomass system to produce its own electricity. They heat manure from their 1,500 cows to produce methane gas, which is collected and used to power a generator. The owner estimates his cows will eventually supply enough electricity to power more than three hundred households. And if you're worried about the stench, removing the methane from the manure gets rid of most of the odor.

And in Renton, Washington, a suburb of Seattle, some seven hundred thousand residents flush eighty-six million gallons of sewage daily into the sewer system. Approximately thirty million gallons of this sewage is sent to tanks, where bacteria convert much of the waste to methane gas. The methane gas is broken down into hydrogen and carbon dioxide. The hydrogen is used as fuel in a one-megawatt fuel-cell power plant, capable of generating enough electricity for one thousand homes.

Methane gas can also be generated from landfills using a technology developed by Viktor Popov at the West Institute of Technology. The landfill is covered with a membrane—a middle permeable layer between two impermeable layers—to prevent air from contaminating the methane. Carbon dioxide extracted from the landfill is pumped into the permeable layer, where it is slightly above atmospheric pressure, preventing air from being drawn into the landfill. As methane is pumped out of the ground, carbon dioxide is drawn through a pipe into the landfill from the membrane.

Fueling the search for alternate energy is the belief that these sources are cleaner than burning coal, oil, and other hydrocarbons. But that's not always the case. Hydroelectric dams can in some cases produce more greenhouses gases—carbon dioxide and methane—than conventional power plants running on fossil fuels. When reservoirs are initially flooded, plants are covered with water. The plants decompose, resulting in a build-up of dissolved methane, which is released into the atmosphere when the water is cycled through the hydroelectric plant's turbine.

Even windmills are not without problems: a Princeton University study has found that large wind farms with ten thousand turbines can actually produce an undesired change in local weather patterns, increasing cloud formation and rain.

Alternate Universes

I N "ALTERNATE ENERGY" we talked about Isaac Asimov's novel *The Gods Themselves*, in which an alternate energy source was created when our universe exchanged matter with a parallel universe.

The notion of an alternate universe, also called a parallel world, has its roots in fairy tales and the "astral plane" of spiritualists and mystics. It has been given scientific confirmation in recent years with speculations about the Big Bang, black holes (and universes possibly budding from black holes), and quantum theory, including Heisenberg's Uncertainty Principle.

The classic alternate universe book is Edwin Abbott's 1884 story "Flatland," in which a universe exists that has only two dimensions. Abbott wrote the book to explain mathematics in an entertaining fashion.

A number of early science fiction and fantasy stories deal with alternate worlds. These include:

- William Hope Hodgson's *The House on the Borderland* (1908) and *The Ghost Pirate* (1909).
- Homer Eon Flint's and Austin Hall's *The Blind Spot* (1921).
- Edmond Hamilton's "Locked Worlds" (1929).
- Murray Leinster's "The Fifth-Dimension Catapult" (1931).
- Clifford Simak's *Ring around the Sun* (1953).
- Keith Laumer's *Worlds of the Imperium* (1962).
- Bob Shaw's *A Wreath of Stars* (1976).
- Roger Zelazny's *Nine Princes in Amber* (1970).
- Frederik Pohl's *The Coming of the Quantum Cats* (1986).

In Jane Lindskold's 1996 novel *Chronomaster*, the protagonist, Rene Korda, is one of a handful of specialists who can create "pocket universes"—small-scale alternate universes located within our own universe:

Korda had begun as a terraformer—an artist in the sculpting of planets not merely to make them habitable, but to fulfill the whims and wishes of his clients. From that it had been a logical development to learn how to create pocket universes.

The training had taken half a century, but he had considered the decades well spent. Afterward, he had created a handful of pocket universes, crafting every detail, from the essential laws of physics to the fauna and flora on each planet. A pocket universe—not much larger than a solar system, but for a human the sensation of creating one was godlike.

Many scientists and mathematicians, including Martin Gardner and Paul Davies, believe the idea of alternate universes to be science fiction inventions—unlikely to be proven a reality by science. Gardner calls alternate universes "frivolous fantasies."

Cosmologists Max Tegmark, Andrei Linde, and other scientists, however, argue that alternate universes could indeed exist.

One argument in favor of alternate universes says that there is no reason to think that the Big Bang was a unique event. If other big bangs—like the one that formed our own universe—take place occasionally, at vast distances throughout an infinite space, then each might have created its own universe.

The universe we can actually observe is about fourteen billion light-years long. What if this "reality" of ours is only a small patch in an infinite space filled with other universes, each created by its own big bang? The laws of physics in the alternate universes may be different than our own. Linde theorizes that wormholes may link these universes together. If these alternate universes spring up often enough, the pattern of matter in our universe may be randomly replicated in one of these alternate universes. Such a universe would be a "parallel" universe in that it is a replica of our own.

But what are the chances that our own universe could be replicated precisely? Well, the arrangement of particles in any universe can be expressed as data, or information, describing the position and characteristics of each particle. So the question is: what are the chances of our particular arrangement being duplicated elsewhere? Answer: If space is infinite, this arrangement or pattern is bound to repeat at some point (in the same way that, if you randomly hit the keys on your PC an infinite amount of times, you will eventually type the pattern of letters and spaces in Shakespeare's plays).

Most of the alternate universes, however, would probably have little resemblance to ours. Even the laws of physics, such as electromagnetism and gravity, could vary greatly from universe to universe. These duplicate universes could even have more dimensions than our own.

Using calculations I can't even pretend to understand, Tegmark estimates that this parallel universe, which would include a duplicate of the entire planet Earth as well as you and me individually, is located within a distance of 10 to the 10^{28} meters from here. Since the distance is far greater than you could ever hope to travel, even at light speed you're probably not in much danger of running into your alternate-universe twin.

Another theory, known as the "M-theory," says that before the Big Bang that formed our universe, the cosmos consisted of two flat four-dimensional surfaces or "sheets." One of these sheets is the space occupied by our universe. The other is a hidden parallel universe. Random fluctuations in this parallel universe cause it to distort and reach toward our universe; portions of the parallel universe occasionally spill over into our own.

Science fiction author Larry Niven has speculated that parallel universes exist apart from our own in another dimension, not another location in space. He whimsically imagines that perhaps fog is caused not by condensed water vapor, but when two parallel universes intersect in the same time and space. If you wander out on a foggy night, you may end up in a world that is not quite your own, never able to find your way back. Niven's idea might not be just fantasy, however. Some cosmologists speculate that alternate universes may exist in a fifth dimension (or even a sixth or higher) beyond the four dimensions (length, width, height, and time).

Another possibility: miniature alternate universes existing inside a black hole. The motion picture *Men in Black II* had fun with the idea of a miniature alternate universe, hinting that our entire universe resides in a locker located in a train station within another universe.

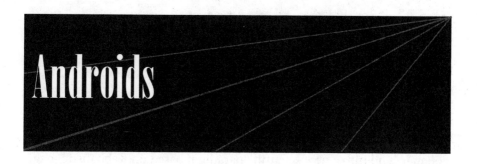

Androids

UNLIKE A ROBOT, which is built from metal and electronic components and is clearly a machine, an android is an artificial being that looks like a man. Androids are made either entirely or partially from biological components— or components designed to simulate the function and appearance of human organs. If an artificial being has no biological components and is totally mechanical, then he's a robot, not an android.

The term "android" was first used in 1727 to describe the attempts of thirteenth-century alchemist Albertus Magnus to create an artificial man. The first science fiction story that used the term "androids" seems to have been in Jack Williamson's novel *The Cometeers* (1936).

Although he didn't use the term, the robots in Karel Capek's 1920 play *R.U.R.* (*Rossum's Universal Robots*) were in fact androids. Capek described his "robots" as "an artificial human of organic substance," which makes them androids, not robots. In that sense, Mary Shelley's Frankenstein's monster could be considered an android.

Edgar Rice Burroughs referred to synthetic men in *The Monster Men* (1929) and *Synthetic Men of Mars* (1940). Edward Hamilton popularized an android in his Captain Future stories (1940s).

In 1976, Isaac Asimov published his story "The Bicentennial Man," about a robot, Andrew Martin, who gradually replaces his robotic parts with components more closely resembling human organs, and in so doing transforms himself from a robot into an android. Andrew describes one of the modifications he has planned for his body:

> "I am designing a system for allowing androids—myself—to gain energy from the combustion of hydrocarbons, rather than from atomic cells. I think I have designed an adequate combustion chamber for catalyzed controlled breakdown."
>
> Paul raised his eyebrows. "So that they will breathe and eat? But why, Andrew? The atomic cell is surely infinitely better."
>
> "In some ways, perhaps, but the atomic cell is inhuman."

The Hebrew golem is the earliest android to appear in fiction, although strictly speaking, he did not contain biological components; he was made of mud and animated through prayer—specifically, through a prayer written on parchment stuffed into his mouth. That parchment may have been the origin of the notion of a machine being controlled by a set of written instructions—a "program." Ted Chiang rewrote the golem legend into a new scientific paradigm based on Kabbalistic magic in "Seventy-Two Letters" (2000).

Clifford Simak wrote a series of stories in the 1950s about androids that were treated as slaves by their human creators. Robert Silverberg wrote about androids in *Tower of Glass* (1970) and C. J. Cherryh in *Port Eternity* (1982).

One of the most famous science fiction stories about androids is the 1968 Philip K. Dick novel, *Do Androids Dream of Electric Sheep?*, in which incredibly humanlike androids, banned from Earth, fight back against their creators. The book formed the basis of the film *Blade Runner* starring Harrison Ford.

As for the most famous android in science fiction, that honor goes to Lieutenant Commander Data of *Star Trek: The Next Generation*. He is the first android to be admitted to Starfleet, although with his electrical system he is part robot. Data serves under Captain Jean-Luc Picard aboard the *Enterprise-D*. Data was built by Dr. Noonien Soong, Earth's foremost robotic scientist. In some sf stories, such as *Blade Runner*, androids rebel against their human creators. But instead of turning against humankind, Data becomes fascinated with the human race that created him, and actively seeks to become more like them—much like the robot Andrew Martin in Isaac Asimov's "The Bicentennial Man" (1976).

The earliest TV android may be Andromeda, an artificial female lifeform featured in the 1961 British science fiction series *Andromeda*. She was built by a supercomputer which itself was constructed according to instructions transmitted to Earth from the Andromeda Galaxy. Astronomer and science fiction writer Fred Hoyle was the cocreator of the series.

In his 1972 story "The Steam-Driven Boy," John Sladek proposes a method for constructing an android:

[It] had organs analogous to those of a living being. The heart and veins were really an intricate hydraulic system; the liver a tiny distillery to volatilize eaten food and extract oil from it. Part of this oil replenished the veins, part was burned to feed the spleen's miniature steam engine. From this, belts supplied power to the limbs.

Grafton had developed a peculiar substance, "graftonite." A plate of it would vary in thickness according to the intensity of light striking it…used (along with mechanical irises and gelatin lenses) to form the eyes. When a tiny image had been focused on each graftonite retina, a pantographic scriber traced swiftly over it, translating the image to motions in the brain.

Similar levers conveyed motions from the gramophone ears, and from hundreds of tiny pistons all over the body—the sense of touch.

The hydraulic fluid was a suspension of red particles like blood corpuscles. When it oozed to the surface through pores, these were filtered out—it doubled as perspiration.

The brain contained a number of springs, wound to various tensions. With the clockwork connecting them to various limbs, organs and facial features, these comprised [the] "memory."

No one is building androids today or thinking of doing so; *robotics* is the field in which scientists are aggressively turning science fiction into science fact. Synthetic beings of the future will likely be robots controlled by AI (artificial intelligence) microprocessors.

Some robots, like the prototypes of Honda's Asimo, are modeled vaguely after the human body (head, torso, arms, and legs), but are not designed to closely resemble people—although technologically, creating such an android would be possible, using the same techniques with which Disney built the automaton of Abraham Lincoln featured in the 1964 World's Fair (it is still displayed today at Disney World).

Yet the android vision—a synthetic being with both machine and organic parts—may yet be realized. Scientists are now developing "biological computers" incorporating human neurons and other organic components. It's conceivable—though not on the immediate horizon—that an AI biological computer could be used as the "brain" of a mechanical man.

Antigravity

CIENCE FICTION WRITERS imagined antigravity as early as the nineteenth century: Percy Greg's *Across the Zodiac* (1880), C. C. Dail's *Willmoth the Wanderer* (1890), and John Jacob Astor's *A Journey in Other Worlds* (1894).

Perhaps the first science fiction novel to feature antigravity as its central theme was *The First Men in the Moon*, written by H. G. Wells in 1901. Following is his description of the Cavorite, the antigravity material the protagonist uses to build a spaceship in which he travels to the Moon:

> The stuff is opaque to gravitation. It cuts off objects from gravitating towards each other.
>
> As soon as it reached a temperature of sixty degrees Fahrenheit and the process of its manufacture was complete, the air above it, the portions of roof and ceiling and floor above it, ceased to have weight. I suppose you know—everybody knows nowadays—that, as a usual thing, the air has weight, that it presses on everything at the surface of the earth, presses in all directions, with a pressure of fourteen and a half pounds to the square inch.
>
> You see, over our Cavorite this ceased to be the case, the air there ceased to exert any pressure, and the air round it, and not over the Cavorite, was exerting a pressure of fourteen pounds and a half to the square inch upon this suddenly weightless air.
>
> The air all about the Cavorite crushed in upon the air above it with irresistible force. The air above the Cavorite was forced upwards violently, the air that rushed in to replace it immediately lost weight, ceased to exert any pressure, followed suit, blew the ceiling through and the roof off. It formed a sort of atmospheric fountain, a kind of chimney in the atmosphere.

In his 1955 story "What Goes Up," Arthur C. Clarke describes his own version of a Cavorite-type antigravity field generated accidentally by a nuclear power plant:

As I already said, weight implies energy—lots of it. That energy is entirely due to Earth's gravity field. If you remove an object's weight, that's precisely equivalent to taking it clear outside Earth's gravity.

To take an object clear away from the Earth requires as much work as lifting it four thousand miles against the steady drag of normal gravity. Now the matter zone of force was still on the Earth's surface, but it was weightless. From the energy point of view, therefore, it was outside the Earth's gravity field. It was inaccessible as if it was on top of a four-thousand-mile high mountain.

John W. Campbell's "Who Goes There?" (1938) deals with an alien who builds an antigravity personal jet pack; the story was later made into the horror/science fiction film *The Thing*:

Like a knapsack made of flattened coffee tins, with dangling cloth straps and leather belts, the mechanism clung to the ceiling. A tiny, glaring heart of supernal flame burned in it, yet burned through the ceiling's wood without scorching it. Barclay walked over to it, grasped two of the dangling straps in his hands, and pulled down with an effort. He strapped it about his body. A slight jump carried him in a weirdly slow arc across the room.

They had coffee tins and radio parts, and glass and the machine shop at night. And a week—a whole week—all to itself...antigravity powered by the atomic energy of matter.

In his 1970 novel *Cities in Flight*, James Blish envisions our descendants leaving Earth to travel through space, not in rocket ships or space stations, but in the actual cities of Earth—New York, Los Angeles, Boston, San Francisco. The cities are launched into space using *spin dizzies*—giant antigravity devices that generate a force field around a city and turn it into a self-contained, self-propelled spaceship:

The figures showed that Dirac was right. They also show that Blackett was right. Both magnetism and gravity are phenomena of rotation.

There's a drive-generator on board this ship—the Dillon-Wagoner gravitron polarity generator. The techies have already nicknamed it the spindizzy, because of what it does to the magnetic movement of any atom within its field.

While it's in operation, it absolutely refuses to notice any atom outside its own influence. Furthermore, it will notice no other strain or influence which holds good beyond the borders of that field. It has to be stopped down to almost nothing when it's brought close to a planet, or it won't let you land.

But in deep space…well, it's impervious to meteors and such trash, of course; it's impervious to gravity; and it hasn't the faintest interest in any legislation about top speed limits. It moves in its own continuum, not in the general frame.

This ship came to Ganymede directly from Earth. It did it in a little under two hours, counting maneuvering time. That means that most of the way we made about 55,000 miles per second—with the spindizzy drawing less than five watts of power out of three ordinary No. 6 dry cells.

Arthur C. Clarke describes an antigravity device in his 1966 short story "The Cruel Sky":

Even though he had known about them for two years, and understood something of their basic theory, the Elwin Levitators—or "Levvies" as they had been christened at the lab—still seemed like magic.

Their power-packs stored enough electrical energy to lift a two-hundred-and-fifty-pound weight through a vertical distance of ten miles, which gave an ample safety factor for this mission. The lift-and-descend cycle could be repeated almost indefinitely as the units reacted against the Earth's gravitational field.

On the way up, the battery discharged; on the way down, it was charged again. Since no mechanical process is completely efficient, there was slight loss of energy on each cycle, but it could be repeated at least a hundred times before the units were exhausted.

And in his 1971 short story "Something Wild is Loose," Robert Silverberg describes a rocket ship propelled by an antigravity device called a "gravity drinker" which "spins on its axes, gobbling inertia and pushing up the acceleration."

Science has not yet invented a working antigravity device, but in theory, is it even possible? To answer that question, we have to take a look at the basic nature of gravity.

Everybody living on this planet is familiar with the effects of gravity. You drop an apple, and it falls toward the ground. We know gravity is an attractive force. Now, if we drop this apple in a tub of water, we see that the water is disturbed and that waves are generated along its surface. Most are familiar with the phenomenon of waves, but but few ever think of gravity in those terms. In fact, gravity is manifested as a wave, as well as a force.

Gravity waves are very similar to electromagnetic waves, which propogate outward with the velocity of light (186,000 miles per second). Electromagnetic waves carry energy and momentum which can be transferred to objects in their path. Radio waves, for example, transmit energy that is converted into sound waves when it crosses the path of a radio.

Gravity waves are produced from the acceleration of two or more masses. This acceleration produces a wave, which propagates outward with a velocity equal to that of light. Only one *body* is needed to produce a gravity wave, as long as some portion of the mass distribution is accelerating with respect to the center of mass of the body (as in any rotating non-homogenous mass).

This wave, like its electromagnetic counterpart, does not require a medium in which to travel: gravity waves can propagate in a vacuum. The gravity wave carries energy, which it can transmit to bodies in its path. This results in motion of the bodies bombarded by these waves.

Now, the effects of these waves have probably never been observed by anyone on this planet, even though it is in constant motion. Because of its spherical shape the Earth itself does not generate any gravity waves (the symmetry of the sphere makes it impossible to produce gravity waves). The Earth is very nearly transparent to gravity waves; they pass right through it as if it weren't there at all, like sunlight through a windowpane.

In general, gravitational radiation—no matter what the source—is of very small magnitude. Scientists are currently attempting to detect gravity waves in bursts from exploding stars. In these explosions, much of the star's mass is converted to energy and radiated as gravity waves. Experimentation and a failure to detect the waves has indicated to scientists that the waves are much weaker than originally aniticipated. For gravity waves to be detected by existing measurement devices, a minimum of fifty to one hundred solar masses daily would have to be converted to energy. A solar mass is roughly equivalent to three thousand trillion tons of matter, and the energy equivalent of this mass can be calculated from the well-known equation $E = mc^2$, where E is energy, m is mass and c is the speed of light.

Just as a mass is required to generate gravity waves, so is a mass needed to detect these waves. At the University of Rochester, physicist David Douglass used a large aluminum cylinder as this second mass. If a gravity wave passes through the cylinder, the ends should oscillate with respect to the middle.

The wave's amplitude is described in terms of the ratio of the displacement of one of the cylinder's ends to the length of the cylinder. This ratio is known in physics as the *strain*; hence the gravitational field is a strain field. The detector is sensitive to a displacement of about a hundredth of the diameter of an atom's nucleus.

Says Douglass, "You might think of a gravity wave detector as a very sensitive bell just waiting to be rung.... Although a gravity wave won't interact very long with that bell, the desirable property would be to have the bell ring, or oscillate for hours or days."

In three years of research (1976 to 1979) Douglass found no gravity waves. He says, though, that gravitational radiation is ". . . just part of physics that should exist." Douglass is working with a new and improved gravity wave detector that is about a million times more sensitive than the old device, and should be able to show positive evidence of gravitational radiation. This new detector uses artificially grown sapphires that, unlike natural ones, are very large and, more importantly, flawless. When and if they interact with a gravity wave, they should oscillate for a long period of time, perhaps for days, giving scientists a better chance to study the phenomenon. This is probably the most unusual use found for these costly artificial gems yet.

Silicon may also be used for such single-crystal detectors. Silicon is the material used in integrated circuits in such devices as pocket calculators and transistor radios. The silicon crystals are in the shape of cylinders, with diameters of up to six inches.

The world's largest gravity wave detector, LIGO, consists of two L-shaped detectors. One is in Louisiana, and the second is three thousand kilometers away in Washington. Each detector has vacuum tubes four kilometers long. Laser beams are bounced between mirrors suspended at the ends and the intersection of the L-shaped tubes. When the light beams collide, they produce bands of light and dark stripes. If a gravity wave of sufficient strength passes through the detector, it will change the length of the tubes slightly—less than 10^{-19} meters. When that happens, it will produce a shift in the bands of stripes. That tiny amount of change in such a large structure is proportional to the width of an atom in the space between Earth and Jupiter.

The concept of gravity waves has been with us since Einstein. If gravity is indeed a wave, an antigravity device would need to incorporate a technology that dampened or weakened that wave—or cancelled it altogether. When we figure out how to stop or dampen the wave, we'll be able to stop gravity. And then the sky's the limit—literally.

The new string theory may also permit creation of an antigravity technology. String theory, which posits that all matter and forces consists of vibrating, one-dimensional loops of energy, or "strings," also proposes ten dimensions instead of the familiar four we know (three of space and one of time).

According to a recent article in *Scientific American* (February 2004), ordinary matter cannot escape into these extra dimensions. But gravity can. This leakage might warp the space-time continuum, and if gravity is leaking out of our dimension, it would reduce the gravitational force in our universe. As a result, cosmic expansion might accelerate, and planetary motions could also be affected.

Another scientific possibility for antigravity is "negative mass." Hermann Bondi, a physicist at the University of Cambridge, proposes that every positive mass could be coupled with a negative mass, just as every magnet has both a north and south pole.

The theory of relativity has always assumed the universe to be made up of four dimensions—three of space and one of time. Now, however, string theory suggests that there are more than four dimensions, as many as eleven. When physicist Paul Wesson and his colleagues at the Guelph-Waterloo Physics Institute recalculated the equations of general relativity using an extra dimension—five instead of four—they found the beginning of a theoretical basis allowing for the possibility of a negative mass. In 2011, they will participate in the launch of an experimental satellite designed to detect negative mass.

Antimatter

S ITS NAME IMPLIES, antimatter is the opposite of regular matter. The positron, for example, is a positively charged equivalent of the negatively charged electron.

When antimatter comes in contact with regular matter, it explodes. The energy produced by a matter/antimatter explosion is proportional to the weight of the material, as dictated by Einstein's famous equation $E = mc^2$. Matter in nuclear weapons is converted into energy according to the same equation.

Because the speed of light is such a large number, multiplying the mass by the speed of light squared means an enormous amount of energy is generated from even a tiny amount of matter. For instance, the energy in five grams of antimatter (the mass of a nickel) is equivalent to five Saturn V rockets. A metric ton of antimatter—about the mass of a car—could produce all of the world's electricity for one year.

Antimatter's existence was first proposed in 1928 by physicist Paul Dirac. He believed that for every particle that exists there is a corresponding antiparticle, exactly matching the particle but with an opposite charge. For every electron with a negative charge, then, there is a corresponding antielectron with a positive charge—a positron. For every proton with a positive charge, there is an antiproton with a negative charge. Neutrons have no charge, so the opposite of a neutron, the antineutron, also has no charge—but the neutron and antineutron have magnetic moments opposite in signs relative to their spins.

Carl Anderson discovered positrons in 1932 while studying showers of cosmic particles in a cloud chamber. He saw tracks left by a particle with the exact same mass as an electron, but with a positive charge, which turned out to be the antimatter Dirac predicted. In 1955, antiprotons and antineutrons were observed in particle accelerators by a team lead by Emilio Segre. The discovery made the front page of the *New York Times*.

The antimatter phenomenon soon captured the imagination of science fiction writers. The first science fiction story to deal with antimatter was probably "Minus Planet" by John D. Clark, published in the April 1937 issue of *Astounding Stories*. Antimatter also played a key role in the "seetee" stories published in the 1940s by Jack Williamson.

In 1943, A. E. van Vogt wrote "Storm," in which a huge storm in space is caused by a gas cloud of ordinary matter coming into contact with an antimatter gas cloud.

In 1950, J. Bridger wrote the story "I Am a Stranger Here Myself," in which humankind learns from aliens how to travel faster than the speed of light using a "multiphase travel" technology based on transforming matter into antimatter—a precursor of the U.S.S. *Enterprise's* warp engines.

In Ian Watson's 1975 novel *The Jonah Kit* a scientist named Paul Hammond discovers that the real universe is actually formed out of antimatter, and that the matter universe in which we live is only a copy of the real structure, which will eventually collapse and disappear. Antimatter was also used in Paul Davies' *Fireball* (1987) and in Dan Brown's *Angels and Demons* (2000).

As every sf fan knows, antimatter is a key element in the warp propulsion systems of starships in *Star Trek*. Controlled contact between matter and antimatter in the ship's nacelles (the long cylinders connected to the main hull) generates the power needed to reach warp (faster than light) speeds. The antimatter, frozen antihydrogen, is contained within magnetic fields and not allowed to touch normal matter until it's needed for fuel. James Gunn also uses antimatter as a fuel source in his novel *Gift from the Stars* (2005).

The 1980s U.S. Strategic Defense Initiative (also known as "Star Wars") explored the possible use of antimatter as a rocket fuel or spaceborne weapon.

Today, Gene Roddenberry's fictitious antihydrogen is becoming a reality. At CERN, the world's largest particle physics center, an antiproton decelerator is being used to produce low-energy antiprotons for the synthesis of antihydrogen—the creation of antimatter. The antiprotons are formed when subatomic particles within an accelerator are made to collide at high speed. However, because linear accelerators require a huge amount of energy to get particles to move at high speeds, the production of antiprotons is grossly inefficient. The amount of antiprotons produced at CERN in one year would supply enough electricity to light a 100-watt electric bulb for just three seconds.

Artificial Intelligence

ARTIFICIAL INTELLIGENCE (AI) refers to a computer or robot whose central processing unit is so powerful, fast, and sophisticated that it can mimic human intelligence, thinking, and possibly even emotions. Many sf writers have expressed concern that an advanced AI computer may even become self-aware.

AI is one of science fiction's longest-running and most popular themes:

- In Harlan Ellison's classic short story "I Have No Mouth and I Must Scream" (1967), a network of AI computers that collectively calls itself "AM" destroys the human race and gives the only five surviving humans greatly extended life spans—so it can hold them prisoners and torture them for all eternity.
- In the 1970 movie *Colossus: The Forbin Project* (based on D. F. Jones' 1966 novel of the same title), an American supercomputer becomes self-aware and takes control of Earth's weapons systems away from the humans who created it.
- In the film *2001: A Space Odyssey* (1968), the supercomputer HAL takes control of the spaceship sent to investigate the monolith and even kills one of the crew members.
- In the 1973 movie *Westworld*, an intelligent android played by Yul Brynner begins killing humans who use him as an amusement at a robotic theme park.
- In the Terminator films, the supercomputer network Skynet becomes self-aware and, a minute later, launches a nuclear attack that kills most of the human population. It then builds robots ("terminators") to hunt down and kill the remaining human population, believing humankind to be a threat to Skynet's existence.
- In his story "The Bicentennial Man" (1976), Isaac Asimov's main character is a robot, Andrew Martin, who has creativity and inde-

pendent thought equal or superior to a human's, and undergoes a series of "upgrades" designed to turn him into a man (see chapter on "Androids").

- In the Matrix movies, a self-aware supercomputer rules the planet after nuclear war causes a nuclear winter that makes the planet's surface uninhabitable, forcing the surviving humans to live deep underground.
- In Steven Spielberg's film *AI*, a robotics company builds an advanced robot that perfectly mimics the intelligence and emotions of a young boy. The film is based on Brian Aldiss' short story "Super-Toys Last All Summer Long," published in 1969.
- In Vernor Vinge's *A Fire Upon the Deep* (1992), AIs begin to compete with human beings for dominance of the galaxy.
- And in his Great Sky River series, Gregory Benford has humanity clinging to survival in a galaxy ruled by AIs.

Both *The Terminator* and *The Matrix* revolve around a sophisticated computer network that evolves in both speed and size until it becomes self-aware. Arthur C. Clarke suggested that this could also happen to the world's telecommunications networks in his 1963 story "Dial F for Frankenstein":

You're right about the fifteen billion neurons in the human brain. Fifteen billion sounds a large number, but it isn't. Round about the 1960s, there were more than that number of individual switches in the world's autoexchanges. Today, there are approximately five times as many.

And as from yesterday, they've all become capable of full interconnection, now that the satellite links have gone into service.

Until today, they've been largely independent, autonomous. But now we've suddenly multiplied the connecting links, the networks have all merged together, and we've reached criticality.

Radio and TV stations [will] be feeding information into it, through their landlines. That should give it something to think about!

Then there would be all the data stored in all the computers; it would have access to that—and to the electronic libraries, the radar tracking systems, the telemarketing in the automatic factories. Oh, it would have enough sense organs! We can't begin to imagine its picture of the world; but it would be infinitely richer and more complex than ours.

As microprocessors become faster and perform at unprecedented levels—Intel's Pentium 4 processor, for instance, operates at 2.2 billion cycles per second—computers and robots increasingly gain the ability to outthink and outperform their human creators.

NASA is developing a supercomputer that will consist of twenty Silicon Graphic servers containing hundreds of Pentium processors. The NASA system may reach a processing speed of sixty teraflops—or sixty trillion "floating point operations" (calculations) per second.

The current speed champion, IBM's BlueGene/L supercomputer, is capable of processing data at a top speed of ninety-two trillion floating point operations per second (that record will be obsolete by the time this book is printed). But that's a snail's pace compared with the "penta-flop computer" IBM is developing: a supercomputer capable of performing 1,000 trillion computations per second.

In his 1961 book *Thinking Machines*, Irving Adler considers the question of whether a computer processor could be fast enough and powerful to duplicate the thought processes of a human brain:

> How do the brain and the machine compare in complexity and efficiency as pieces of a thinking apparatus? We can get a crude answer to this question from certain known facts about the brain and machines.
>
> We can use as an index of relative complexity the ratio of the number of unit actions that can be performed by equal volumes of each in equal times. This can be calculated from the ratio of the volumes of the basic units in each, and the ratio of their speeds.
>
> A neuron is about one billion times smaller that the basic unit of a calculating machine, and it is about one hundred thousand times slower in its action. Dividing these two figures gives us the desired index. Nerve tissue is about ten thousand times more efficient than electronic hardware.
>
> An aspect of the brain's greater complexity is the great size of its memory capacity. A large modern computer has a memory capacity of one million bits. The memory capacity of the brain has been estimated at 280 billion billion bits. So the brain's memory capacity is 280 million million times as great as that of any existing machine.

Adler wrote this in 1961, way before the introduction of the personal computer. The first PCs, first marketed to the general public in the early 1980s, had just 640,000 bytes of memory. Two decades later, computer systems for beginners are packing 512 million bytes of RAM, and many users are upgrading to two billion and even four billion bytes of memory. And we always want more.

Roger Zelazny, in his 1975 novella "Home is the Hangman," describes a robot brain which achieves artificial intelligence because it has approximately the same number of computing elements as there are neurons in the human brain:

> Last century, three engineers at the University of Wisconsin—Nordman, Parmentier and Scott—developed a device known as a superconductive tunnel-junction neuristor. Two tiny strips of metal with a thin insulating layer between. Supercool it and it passed electrical impulses without resistance. Surround it with magnetized material and pack a mass of them together—billions—and what have you got?
>
> Well, for one thing you've got an impossible situation to schematize when considering all the paths and interconnections that may be formed. There is an obvious similarity to the structure of the brain.
>
> So, they theorized, you don't even attempt to hook up such a device. You pulse in data and let it establish its own preferential pathways, by means of the magnetic material's becoming increasingly magnetized each time the current passes through it, thus cutting the resistance. The material establishes its own routes in a fashion analogous to the functioning of the brain when it is learning something.
>
> In the case of the Hangman, they used a setup very similar to this and they were able to pack over ten billion neuristor-type cells into a very small area—around a cubic foot. They aimed for that magic figure because that is approximately the number of nerve cells in the human brain. That is what I meant when I said that it wasn't really a computer. They were actually working in the area of artificial intelligence, no matter what they called it.

Do computers already out-think people? Let's consider Deep Blue, the IBM supercomputer that defeated world champion Garry Kasparov at chess. The defeat of a human grandmaster by an intelligent supercomputer was predicted decades ago by Fritz Leiber in his short story "The 64-Square Madhouse" (1962). Deep Blue—an IBM RS/6000 supercomputer with thirty-two processors—can consider two hundred million moves per second. Kasparov believes the computer can actually think. He says, "The computer is very human." And Deep Blue has not nearly approached the limit of processing power in a high-speed computer: by adding another 480 processors, IBM engineers say they could increase the capacity to two billion moves per second.

In his 1950 article "Computing Machinery and Intelligence," Alan Turing proposed what is known as the "Turing Test," arguing that if a machine could successfully pretend to be human to a knowledgeable observer, then you certainly should consider it intelligent. By some standards (although not necessarily the Turing Test), IBM's Deep Blue is a working example of AI. Maurice Ashley, a top American chess player who provided commentary on the match between Kasparov and Deep Blue, said, "Deep Blue has shown human tenacity."

Arthur C. Clarke says that the creation of ultra-intelligent AI machines is just a few years away, and that when they are built, they will in essence be a new species. Inventor and author Ray Kurzweil predicts that computers will match the computational functions of the human brain by 2030, and that soon afterward humans and computers will merge to become a new species.

The latest development on the AI front is "genetic programming"— computer programs that "evolve" to achieve machine intelligence. Genetic programming (GP) is an automated method for creating a working computer program from a high-level statement of a problem. Genetic programming starts from a high-level statement of "what needs to be done" and automatically creates a computer program to solve the problem. This "genetic software" is able to create inventions automatically, just like a human inventor. In fact, such programs have already duplicated fifteen previously patented electronic inventions.

A 2003 article in *Scientific American* observes that genetic programs are now routinely replicating human inventions, just half a century after Alan Turing predicted that "human-competitive machines" would be built. It goes on to say, "The goal of artificial intelligence and machine learning is to get computers to solve problems from a high-level statement of what needs to be done."

One way scientists hope to give computers the ability to "think" rather than just crunch numbers is by integrating biological components and machines. U.S. scientists recently built a computer made of neurons taken from leeches. The aim is to develop a new generation of "biological computers" that can figure out how to solve problems with little or no operator intervention—in other words, to think for themselves. In the first generation biological computer, built by Professor Bill Ditto at Georgia Institute of Technology, the leech neurons are placed in a Petri dish and a micro-electrode is inserted into each neuron. The neurons have their own electrical activity and respond to stimulation from the electrode. The electric impulses from the electrode can be used to make each neuron represent a binary number, 0 or 1. Conventional computers also process information as binary patterns. Ditto envisions building intelligent robots whose brains are intelligent biological computers—a combination of electrodes and neurons.

A different type of biological computer was built by USC science professor Dr. Leonard Adleman. Unlike Ditto's machine, which uses neurons to perform calculations, the biological component of Aldeman's computer is DNA. In Aldeman's biological computer, a strand of DNA represents a math or logical problem. The computer generates trillions of other unique DNA strands, each being a possible solution to the problem. By monitoring the way the DNA strands bind to one another, the computer can eliminate invalid solutions. After the process of elimination, the DNA strand remaining is the solution to the problem. "These biological molecules are miraculous little machines," says Aldeman. "They store energy and information, and they cut, paste, and copy."

Most computers require relatively cool temperatures to operate. Large data centers are air-conditioned, and all PCs have an internal fan which circulates air inside the housing so the microprocessor does not overheat. But DNA-based biological computers may actually process data more efficiently at high temperatures. Chemical engineers at Seoul National University found that it may be better to operate DNA computers at the melting point of DNA, the temperature at which the double helix of the DNA molecule separates into single strands (95 degrees Centigrade, or 203 degrees Fahrenheit).

In one experiment, the engineers used a DNA computer to solve a classic problem in computer science: calculating the shortest circular route between seven cities in a salesperson's territory. The goal: find the shortest route along which the salesperson could start out from home, visit all cities in the route, and then return home. Each city and route was represented by a unique DNA sequence; the longer the distance, the higher the melting point. DNA strands were mixed together to generate each possible path. Strands starting and ending at the home city were multiplied. Strands not containing all cities on the route were discarded. The remaining strands, all representing viable routes, were measured by applying heat and determining the melting point. The one that melted first (because it had the lowest melting point) was the shortest route. Another nonelectronic computer is a prototype for a "chemical computer" developed by Andrew Adamatzky, a computer scientist at the

University of the West of England. The "central processing unit" of the computer is a dish of chemicals sitting on a lab bench. Computations are performed not by electronic bits and bytes, but by waves of ions that form spontaneously and diffuse through the mixture.

Not every scientist is thrilled by the prospect of AI. Some agree with science fiction writers who predict AI machines taking control of the planet away from humans.

"The danger is real that computer intelligence will take over the world," says acclaimed physicist Stephen Hawking. He predicts the development of direct links between human brains and computers. With a direct interface between neurons and an AI machine, says Hawking, AI computers could contribute to human intelligence rather than oppose it.

If the mark of intelligent machines is the Turing Test—the ability to fool a human observer into thinking the machine is a person—then AI software already exists: on the Internet. According to a 2002 article in the *New York Times*, "Rogue computer programs masquerading as teenagers" have entered chat rooms, collected personal information, and posted links. People running the chat rooms were unable to distinguish postings by the software from postings by real people. Therefore, the software passed the Turing Test with flying colors.

Also passing the Turing Test is a "robot scientist"—really a PC-based bench top system—built by computer science professor Ross King at the University of Wales. Like a human scientist, this PC-based intelligence is capable of generating hypotheses, designing and carrying out experiments to test them, and interpreting the results. In one experiment, a problem in genetics was assigned to both the robot scientist as well as human scientists. In one key measure, the robot solves the problem as adroitly as its flesh-and-blood colleagues.

Another route to artificial intelligence is proposed by physicist David Nolte in his book *Mind at Light Speed* (2001): to mimic or possibly replicate human thought and intelligence, computers need to attain very high processing speeds; IBM's Big Blue is a prime example. Nolte and his colleagues are building these high-speed processors in a new generation of computers, made possible by the discovery of fiber optics: ultrathin optical cables that carry signals as laser light. By using photons (light) instead of electronics (electricity) to represent data, optical computers can process information much faster than electronic computers, based on the assumption that nothing travels faster than light (see "Faster-than-Light Travel" in the appendix). The most powerful of these new AI machines, according to Nolte, will be "quantum optical" computers.

These quantum optical computing machines will use the quantum nature of the photon to represent data as "quantum bits." Since the quanta is the smallest unit of mass and energy, processing of quantum bits (called "qubits") will be more efficient than processing bits represented by larger particles such as photons or electronics.

For instance, a conventional computer using a classical algorithm at terahertz speed would take 150,000 years to calculate the prime factors of a three-hundred-digit number. A quantum-factoring algorithm using qubits could perform the same calculation in less than a second.

Artificial Life

"ARTIFICIAL LIFE SHOULD BE distinguished from robots and androids," says science fiction historian and author James Gunn. "Artificial life refers to everything that is not a robot, android, or artificial intelligence or computer." According to Gunn, the medieval golem could qualify as artificial life, as well as Theodore Sturgeon's microscopic people in "Microcosmic God" (1941) and Greg Bear's intelligent bacteria in *Blood Music* (1985).

In his book *Virtual Organisms* (1999), Mark Ward uses slightly different criteria to define artificial life. He says that artificial life encompasses "software simulations, robotics, protein electronics, and even attempts to recreate the Earth's first living organisms. It is less concerned with what something is built of than with how it lives."

Genetic engineering creates new life-forms, but they are not really "artificial" life, because the altered genes are created from other living genes.

The most famous science fiction story about artificial life is, of course, Mary Shelley's *Frankenstein*, published in 1818. Victor Frankenstein, a scientist interested in alchemy, robs graves and sews parts taken from dead bodies into a new body (no explanation is given as to why he just didn't take a whole dead body and reanimate it). He harnesses the electricity of a lightning bolt to bring life to the inanimate body:

> Although I possessed the capacity of bestowing animation, yet to prepare a frame for the reception of it, with all its intricacies of fibers, muscles, and veins, still remained a work of inconceivable difficulty and labor.
>
> I began the creation of a human being. Life and death appeared to me ideal bounds, which I should first break through, and pour a torrent of light into our dark world. A new species would bless me as its creator and source; many happy and excellent natures would owe their being to me. I thought that if I could bestow animation upon

lifeless matter, I might in process of time (although I now found it impossible) renew life where death had apparently devoted the body to corruption.

I collected bones from charnel-houses and distributed, with profane fingers, the tremendous secrets of the human frame. In a solitary chamber, or rather cell, at the top of the house, and separated from all the other apartments by a gallery and staircase, I kept my workshop of filthy creation; my eyeballs were starting from their sockets in attending to the details of my employment. The dissecting room and the slaughter-house furnished many of my materials; and often did my human nature turn with loathing from my occupation, whilst, still urged on by an eagerness which perpetually increased, I brought my work near to a conclusion.

It was on a dreary night that I beheld the accomplishment of my toils. With an anxiety that almost amounted to agony, I collected the instruments of life around me, that I might infuse a spark of being into the lifeless thing that lay at my feet. My candle was nearly burnt out, when, by the glimmer of the half-extinguished light, I saw the dull yellow eye of the creature open; it breathed hard, and a convulsive motion agitated its limbs.

His limbs were in proportion, and I had selected his features as beautiful. Beautiful! Great God! His yellow skin scarcely covered the work of muscles and arteries beneath; his hair was of a lustrous black, and flowing; his teeth of a pearly whiteness; but these luxuriances only formed a more horrid contrast with his watery eyes, that seemed almost of the same colour as the dun-white sockets in which they were set, his shriveled complexion and straight black lips.

Life has already been created out of nonlife once on Earth. Scientists believe this probably happened over 3.5 billion years ago when organic chemicals in the ocean formed into amino acids, the building blocks of protein. And as with Frankenstein's monster, lightning played a pivotal role, providing the energy needed for the precise chemical reactions leading to the creation of life.

The oldest fossils dating back to the period on Earth when life originated are bacteria called *cyanobacteria*. These are a type of blue-green algae that secreted a thick cell wall. In 1953, Stanley Miller of the University of Chicago attempted to recreate the environmental conditions of primordial Earth that led to the initial formation of amino acids. In a closed flask, Miller placed methane, ammonia, hydrogen, and water, chemicals believed to be the major components of the early Earth's atmosphere. He then ran a continuous electric current through the glass container to simulate the lightning storms that took place frequently on early Earth. After a week, he examined the contents of the flask and found a small quantity of amino acids—the basic building block of living organisms—present.

In 1961, Juan Oro conducted a similar experiment by placing hydrogen cyanide and ammonia in an aqueous solution. The experiment produced adenine, one of the four bases in RNA and DNA.

But, while Miller and Oro produced the precursors of life from chemical compounds, neither they nor anyone since has been able to create an actual living organism in this fashion.

Hamilton Smith, a Nobel laureate in Rockville, Maryland, has announced plans to create a single-celled partially man-made organism in a laboratory. If he is successful, his organism will be true artificial life: the cell will feed and divide to create a population of new cells never before seen on Earth. To create this artificial organism, Smith will remove the genetic material from a microscopic organism, and replace it with artificially created chromosomes.

The creation of artificial life is now becoming a field of its own in the science community. In synthetic biology, scientists alter the DNA of an already existing organism, reengineering it for specific purposes. Although the organism is already living, the result is a new type of organism not produced by nature or evolution—a man-made form of life. For instance, using synthetic biology, you might create an organism that swims through your circulatory system and eats plaque from artery walls, preventing heart disease. Biologists have already created a number of new life-forms. One removes heavy metals from wastewater. Another performs simple binary logic—making it useful for biological computing (see "Artificial Intelligence").

Jack Szostak, a researcher at Harvard Medical School, is conducting experiments aimed at creating an artificial single-celled organism from chemicals. He has already been successful in creating self-replicating RNA molecules from nucleotides, but is still searching for a substance that can hold the RNA inside its artificial cell membrane, which is made of fatty acid bubbles known as vesicles, just as a living cell holds its RNA.

Szostak is also working on the problem of getting the artificial cell to divide. By adding montmorillonite, a clay, to the fatty acids, formation of vesicles was accelerated one hundredfold, growing the vesicle to the size where it can support two complete cells following division. The clay can also get nucleotides to fuse together into RNA.

Steen Rasmussen, a physicist at Los Alamos National Laboratory, is also working on a synthetic life-form—nicknamed the "Bug"—which he plans to build from a molecule largely foreign to natural life-forms: peptide nucleic acid (PNA). PNA uses the same "letters" of the genetic code as DNA—A, T, C, and G—enabling PNA to self-replicate in a manner similar to DNA.

Created from nonliving chemicals in a laboratory, PNA-based artificial organisms—a new form of life—could be custom-built for specific tasks ranging from breaking down toxic compounds and producing useful chemicals like hydrogen fuel, to becoming part of machines that can repair themselves like living beings.

To look at artificial life from a *non*biological perspective, one could argue that a sentient machine—an AI computer (see chapter on Artificial Intelligence)—represents a type of artificial life. Michael Crichton describes this route to artificial life in his novel *Prey* (2002):

> In the last few years, artificial life had replaced artificial intelligence as a long-term computing goal. The idea was to write programs that had the attributes of living creatures—the ability to adapt, cooperate, learn, and adjust to change. Many of those qualities were especially important in robotics, and they were starting to be realized with distributed processing.
>
> Distributed processing meant that you divided your work among several processors, or among a network of virtual agents that you created in the computer. There were several basic ways this was done. One way was to create a large population of fairly dumb agents that worked together to accomplish a goal—just like a colony of ants worked together to accomplish a goal.
>
> Another method was to make a so-called neural network that mimicked the network of neurons in the human brain. It turned out that even simple neural networks had surprising power. These networks could learn.
>
> A third technique was to create virtual genes in the computer and let them evolve in a virtual world until some goal was attained.

The "neurons" in an AI brain do not have to be biological: scientists in Colorado are working to develop a robotic brain with 3.7 million artificial neurons, each consisting of a group of transistors contained in a cell. The researchers are hoping that the artificial brain will allow a robot to interact with stimuli in its environment and develop the sort of intelligence seen in animals.

Sony has created Aibo, an artificial dog with a CPU for a brain and sensors providing the feedback of eyes and ears. Pressure sensors enable Aibo to detect when you are petting him. Eyes with a built-in 180,000-pixel miniature color video camera can recognize objects. Software allows Aibo to perform complex actions, display the semblance of emotions and instincts, and learn through experience, a possible step toward true AI.

An ultra-intelligent AI robot could even build copies of itself, making it self-replicating—a condition basic to living things. Maybe Abio will have puppies. If the artificial dog could replicate itself mechanically, it could download data stored in its CPU to duplicate its "consciousness" and learning to each of its robotic offspring.

Asteroids Colliding with the Earth

PACE IN GENERAL, and our solar system in particular, is filled with traveling objects both small and large, including asteroids and comets. Thousands of them orbit our Sun in the asteroid belt between Mars and Jupiter. Giuseppe Piazzi discovered the first asteroid, which he named Ceres, in 1801. The asteroids Vesta and Pallas were discovered within the next ten years. More than two thousand asteroids have since been cataloged.

A number of early sf writers set their tales on asteroids: Clifford Simak's "The Asteroid of Gold" (1932), Stanton A. Coblentz's "The Golden Planetoid" (1935), and Jack Williamson's *Seetee Ship* (1942).

For some time, sf writers conceived of asteroids as western-like frontiers for explorers, adventurers, and fighting; for instance, Isaac Asimov's *Lucky Starr and the Pirates of the Asteroids* (1953). More recent stories consider the possibility of hollowed-out asteroids as habitats, as in George Zebrowski's *Macrolife* (1979).

Asteroids are not always used in a benign way, however. The idea of cosmic collisions probably began with H. G. Wells' "The Star" (1897). The most famous science fiction novel featuring impact with an asteroid is Edwin Balmer's and Philip Wylie's *When Worlds Collide* (1933).

In the 1950s TV series *Superman,* the Man of Steel flies into outer space to stop an asteroid from crashing into Earth. He plants a bomb on the asteroid and succeeds in blowing it to pieces.

In their 1977 novel *Lucifer's Hammer*, Larry Niven and Jerry Pournelle describe the consequences of a large asteroid or meteor striking the Earth, and suggest that breaking it up into smaller pieces would increase the damage.

In the 1998 movie *Armageddon*, Bruce Willis leads a team of astronauts who must land on a large asteroid headed for Earth and blow it up before it collides with our planet. That same year, the film *Deep Impact*, starring Morgan Freeman and Robert Duvall, also dealt with an asteroid on a collision course with Earth.

Is it inconceivable that an asteroid could crash into Earth with devastating effect? Not at all. Objects from space strike the Earth all the time.

Most are no bigger than a grain of rice, and quickly burn up in our atmosphere as "shooting stars." Some survive and strike the surface, occasionally damaging a home or starting a forest fire. The surviving remnants are the meteorites you see in museums.

According to researchers from Bordeaux University Observatory, a flurry of meteors struck the Earth about fifty million years ago, leaving a field of more than one hundred craters ranging in diameter from sixty-six feet to 1.2 miles. Astronomers at the Free University of Brussels have shown that a massive asteroid struck the Earth about thirty-five million years ago, leaving a one-hundred-kilometer crater in Siberia.

On November 30, 1954, a meteorite weighing nearly eight pounds crashed through the roof of a home in Alabama, knocking a hole in the ceiling and smashing a radio. And in October 1992, a chunk of a meteor weighing almost twenty-eight pounds smashed into a high school student's Chevrolet Malibu, punching a hole through the trunk and leaving a bathtub-size crater in the ground underneath it.

In the past eight years, approximately three hundred asteroids—ranging in size from three to thirty feet wide—exploded in Earth's upper atmosphere, one with five kilotons of force. Such an explosion could be mistaken for a nuclear detonation and possibly trigger retaliation with atomic weapons, starting World War III.

Scientists speculate that a giant asteroid, approximately six miles in diameter, crashed into the Earth sixty-five million years ago and caused the extinction of the dinosaurs. The impact kicked up vast dust clouds that covered most of the planet.

But the largest mass extinction took place 250 million years ago when a meteor the size of Mount Everest crashed into the Earth off the northwestern border of Australia, leaving a crater 125 miles in diameter. The asteroid wiped out seventy percent of land species and ninety percent of marine species. The crash may have set off a worldwide chain of volcanic eruptions.

And just forty-nine thousand years ago, an iron asteroid struck the surface of what is now Arizona, leaving a crater three-quarters of a mile wide, almost certainly killing any living creatures for hundreds of miles around.

In 1908, a meteor estimated to be up to 150 feet wide nearly hit the Earth's surface before burning up over Russia. The explosion was equivalent to ten million tons of TNT—more than 750 times more powerful than the nuclear bomb that exploded over Hiroshima in World War II. It flattened hundreds of square miles of forest in Siberia.

Scientists estimate that an asteroid of this size strikes the Earth once every thousand years.

In 1996, an asteroid a third of a mile wide passed within 280,000 miles of Earth—about the same distance as between the Earth and the Moon. It was the largest object ever to pass that close, and had it hit, it would have caused an explosion in the five thousand- to twelve thousand-megaton range.

In March 2004, a one-hundred-foot-long asteroid passed within 26,500 miles of our planet—the closest a passing asteroid has ever come without actually crashing into us. To give you an idea of how close that really is, that distance is almost the same as the geostationary satellites we place in Earth's orbit to monitor weather. The asteroid was so close, its path was bent fifteen degrees by the pull of our gravity, and it was visible with binoculars.

The threat from this particular asteroid, named 2004 FH, is not over, however. Calculations show that 2004 FH circles the Sun every nine months, and its trajectory always brings it close to the Earth. A collision with 2004 FH or another asteroid is certainly not out of the question. If it came directly toward us, an asteroid the size of 2004 FH would likely break apart or explode in the atmosphere. But if it were just slightly larger, it could survive entry into the atmosphere, hit the surface, and destroy a city-sized area. A larger asteroid, say fifteen to thirty miles wide, could easily wipe out most life on Earth.

In 2002, astronomers discovered an asteroid, labeled 2002 NT7, on an impact course with Earth. Although it's far from certain, the asteroid could strike Earth on February 1, 2019, at a speed of twenty-eight kilometers per second—enough to wipe out a continent and cause a global climate change.

Dr. Benny Peiser of Liverpool's John Moores University says, "This asteroid has now become the most threatening object in the short history of asteroid detection."

Science takes the asteroid threat seriously, stepping up asteroid detection efforts with highly sensitive electronic cameras and automated telescopes that scan the skies for anything that moves in relation to the background stars. China is so concerned about possible collisions with asteroids that its scientists are building an observatory for the sole purpose of detecting near-Earth asteroids (NEAs).

In February 2004, scientists met in California at a Planetary Defense Conference to discuss early detection of asteroids heading for Earth—and to determine how such an asteroid might be deflected. U.S. Representative Dana Rohrabacher of California has proposed an increase in the "planetary defense budget"—funds used for asteroid detection—from $3.5 million to $20 million annually. NASA has set itself a goal of being able to detect ninety percent of the NEAs with a diameter of half a mile or greater.

The key to saving the Earth rests largely on early detection: It takes much less force to deflect the asteroid enough so that it misses Earth while it is still far away. The closer the asteroid, the more energy that is required to prevent the asteroid from striking the planet. Nuclear explosions could be used to either deflect the asteroid or blow it up altogether.

"People talk about blasting an asteroid out of the sky. That's nonsense," writes astronomer Duncan Steel in his book *Target Earth* (2000). "If you break up a near-Earth object heading for Earth, you simply transform a cannonball into a shotgun blast: almost all of the fragments will still hit the planet, causing nearly as much damage as would have been the case anyway." What we need to do, says Steel, is to push the asteroid into a slightly different trajectory that causes it to pass by Earth without harm.

For an asteroid one mile in width, a million tons of TNT would be needed to make an explosion powerful enough to blow it off its direct course to Earth. Since a rocket can't fly a million tons of TNT to a distant asteroid, the only practical alternative is to use a nuclear bomb.

Another asteroid defense system involves traveling to the asteroid in a rocket ship, then assembling a high-powered plasma engine on its surface. In essence, we'd be attaching a rocket engine to the asteroid, and use the engine to drive the asteroid off its collision course with Earth.

Finally, an Earth-based defense system might be used to stop asteroids—for instance, missiles with nuclear warheads. But that might require letting the asteroid get too close for comfort.

Atomic Warfare

$\mathbb{S}$ CIENCE FICTION WRITERS began writing about the destruction of the Earth by an atomic bomb long before one was ever created.

Perhaps the most famous sf use of an atomic bomb to destroy civilization came in H. G. Wells' *The World Set Free* (1914). Physicist Leo Szilard credits Wells' novel with giving him the idea that an atomic bomb was possible, and when he joined the Manhattan Project in 1941 he helped to create one.

In *The Lord of Labour* (1911), George Griffith imagines weapons similar to bazookas firing atomic missiles. And in his book *A Columbus of Space* (1909), Garrett Serviss described a nuclear-powered spaceship. John W. Campbell's first published story, "When the Atoms Failed" (1930), dealt with an atomic bomb. Campbell frequently wrote about atomic weapons in his editorials in *Astounding* and printed them on the covers of the magazine.

Robert Heinlein's "Blowups Happen" (1940) details anxiety in an atomic factory, and Lester del Rey's "Nerves" (1942) describes an accident in an atomic plant.

In his 1941 story "Solution Unsatisfactory," Heinlein suggests that victory in World War II could be achieved by blowing radioactive dust onto the enemy. The U.S. Government has on a few occasions become uncomfortable with the accuracy of several sf writers' descriptions of the atomic bomb. Cleve Cartmill's "Deadline" (1944) describes the bomb so accurately that it prompted the U.S. military to investigate both him and his editor Campbell. The dropping of the atomic bomb did not stop the speculation about a possible atomic catastrophe. Many authors published post-war fiction on the subject:

- Judith Merril's *Shadow on the Hearth* (1950).
- Wilson Tucker's *The Long Loud Silence* (1952).
- Mordecai Roshwald's *Level 7* (1959).
- Pat Frank's *Alas, Babylon* (1959).

- The film *Panic in the Year Zero* (1962).
- Harlan Ellison's "A Boy and His Dog" (1969).
- Roger Zelazny's *Damnation Alley* (1969).

In his 1957 novel *On the Beach*, Nevil Shute depicts the desperate state of a group of post-nuclear-holocaust survivors in Australia. Their world was destroyed when at least 4,700 nuclear weapons were detonated across the globe. Radioactive wind currents are headed toward their area, and the exposure will kill them all. Another post-nuclear-holocaust world is portrayed in the 1979 film *Mad Max* starring Mel Gibson. The few surviving humans either become nomadic loners or band together in small groups. The remaining supplies of gasoline enable transportation by motor vehicle and become the world's most precious and coveted resource.

Walter Miller's classic 1959 novel *A Canticle for Leibowitz* also explores an Earth devastated by nuclear war. Eventually the world is rebuilt, including the nuclear technology that caused the first cataclysm, and humanity once again finds itself on the brink of nuclear war.

The atomic age began in 1919 when Ernest Rutherford bombarded nitrogen gas with helium nuclei (also known as "alpha particles"). By doing so, he became the first human to intentionally produce a nuclear reaction: the helium nucleus and the nitrogen nucleus fused to form a hydrogen nucleus (a proton) and an oxygen nucleus. Rutherford's nuclear reaction carried with it little danger of explosion, as the conversion only took place as long as alpha particles were in contact with the nitrogen gas. A nuclear weapon, however, requires a self-sustaining nuclear reaction: the material being targeted gives off particles as it is transformed; these emitted particles strike other nuclei in the pile. These in turn give off more particles, which strike other nuclei, and so on.

Uranium and plutonium both emit the particles needed to cause a nuclear reaction to take place, making them both radioactive elements. If enough uranium or plutonium is in one place, the particles given off—neutrons—will inevitably strike more of the radioactive material, instead of escaping into the atmosphere; thus there is no need to deliberately aim particles at the target nuclei as Rutherford did. Particles will randomly strike the nuclei, converting them into another element or isotope. In the process, these nuclei give off more particles, which hit more nuclei, and the reaction continues on and on—it becomes a "chain reaction."

An atom bomb contains two pieces of uranium, each of which is individually below the critical mass, the weight at which a self-sustaining chain reaction can take place. A detonation of conventional explosives drives the two pieces of uranium into one another. Together, the combined weight of the two pieces is above the critical mass. Almost instantly, an emitted neutron strikes a uranium or plutonium nucleus, initiating the chain reaction. In less than a second, the material explodes with terrific force.

On July 16, 1945, the first working atomic bomb, built using plutonium, was detonated at a test site in New Mexico—generating an explosion equivalent to the force of twenty thousand tons of TNT. Before the atomic bomb tests, scientists performed calculations to convince themselves that the blast would not ignite the nitrogen in Earth's atmosphere and incinerate the entire planet. The risk was low, and the test went off fine. Two more bombs would later be dropped on Japan, ending World War II.

In a plutonium bomb, such as the one tested in New Mexico, an amount of plutonium below critical mass (less than twenty-two pounds) is surrounded by a heavy layer of inert metal. Conventional explosives are placed around the metal layer, and the inner surface of the bomb's outer shell is coated with beryllium. When the explosives are detonated, the heavy inert metal is driven into the plutonium core, compressing it. The increased density forces the plutonium to reach its critical mass and sets off the chain reaction. The beryllium layer reflects any neutrons traveling away from the plutonium mass back toward the core. The chain reaction accelerates and the bomb explodes.

The chain reaction in an atomic bomb using plutonium or uranium is a fission reaction; when the neutrons hit the plutonium or uranium atoms, they split those atoms apart, driving off neutrons from the nuclei. The opposite process, fusion, is the reaction behind the hydrogen bomb.

In a hydrogen bomb, deuterium and tritium, two isotopes of hydrogen, are fused together to form heavier nuclei. When the hydrogen isotopes fuse to form helium, neutrons are emitted. The neutrons strike more nuclei, causing more isotopes to fuse into helium, and an uncontrolled chain reaction begins. So great is the heat required to fuse the hydrogen into helium that an ordinary atom bomb (a fission bomb) is used as the "trigger" that sets off the fusion reaction. Our Sun uses the same process to convert four million tons of hydrogen to helium every second.

A temperature of many millions of degrees Centigrade is required to sustain a fusion reaction. At these temperatures, the electrons have become so energetic that they are stripped from the atomic nuclei.

In a government program called Project Orion—started in 1957 and shut down in 1964—rocket scientists designed a spaceship for intergalactic travel using hydrogen bombs as the "fuel." A series of hydrogen bombs would be ejected out of the rear of the craft and detonated. The force of the detonation would push against a plate shielding the ship, and launch the ship into outer space at terrific speeds. Larry Niven and Jerry Pournelle described such an atomic-powered ship in their novel *Footfall* (1985).

The Big Bang

F OR MILLENNIA, humankind has wondered about the nature of the
universe, including where it came from and how it got here.

For many years, astronomers supported the "steady state" theory
of the universe, which asserts that the universe has always been here
and always will be: there is no beginning and no end. The universe goes
on forever, basically much as it has always been. Even Einstein supported this theory.

When astronomer Edwin Hubble discovered galaxies beyond the
Milky Way (starting with the Andromeda Galaxy in 1924), he theorized that the universe was not static, but rather dynamic and expanding. Typical expanding universe stories include Olaf Stapledon's *Star
Maker* (1937) and Edmond Hamilton's "The Accursed Galaxy" (1935),
in which other galaxies are fleeing from the Milky Way because of a
disease afflicting our galaxy.

Today the most widely accepted theory of the origin of the universe is
the Big Bang Theory. Georges-Henri Lemaitre suggested the beginnings
of the theory in 1931, and George Gamow added to it later.

David Langford's *Earthdoom* (1987) speculated that the Big Bang
might have resulted from an unwitting time traveler, just as A. E. van
Vogt's "The Seesaw" (1941) suggested that the solar system was created
by a reporter swinging back and forth through time until he explodes
with built-up energy.

In James Blish's *The Triumph of Time* (originally published in 1958 as
A Clash of Cymbals), the Okies witness the end of the universe and the
beginning of another. The same event takes place in Poul Anderson's
Tau Zero (1970).

The Big Bang Theory says that the universe as we know it began as
a tiny mass of "primordial matter." At a point in time that marks the
beginning of time as we know it, the density of the mass caused it to
reach a temperature of ten billion degrees. At that critical temperature,
the mass exploded. In less than one second, the matter began expand-

ing into space, forming the elements we know today, starting with hydrogen. About 4.6 billion years ago, much of the matter in the universe was contained in large clouds of hot gas. According to a new theory by astronomer Jeff Hester of Arizona State University, intense ultraviolet radiation created a bubble within one of these gas clouds. A shock wave compressed the surrounding gas, triggering the formation of many of the low-mass stars in our universe today, including our own Sun. Intense ultraviolet radiation from a nearby massive star evaporated much of the gas surrounding the star, leaving behind a small star and a flat disk of gas and dust. The gas and dusk in the disk formed the planets in our solar system.

Alan Guth of MIT first proposed the "Inflation Theory" to explain why galaxies in our universe are flying apart. According to Inflation Theory, everything in our universe came from one small, hot, dense ball of matter, which blew up at the time of the Big Bang to form everything that exists in our universe today.

Another theory of the universe, known as the "Oscillating Theory," has been proposed by Neil Turok, chair of mathematical physics at Cambridge University. The Oscillating Theory says that the universe is continually expanding and contracting. The universe as we know it, according to this theory, will eventually collapse upon itself, forming a dense ball of matter. The matter will explode in another Big Bang-like event, and the whole thing will start all over again.

When the universe began at the Big Bang, one of two things happened. The first scenario is that the explosion was forceful enough to jettison matter forward at a velocity great enough to escape the gravitational pull of the matter at the center of the universe. In this scenario the universe will continue to expand indefinitely.

The second scenario, which proponents of the Oscillating Theory support, is that the Big Bang explosion was not strong enough for the matter to escape the gravitational pull of matter at the center of the universe. If that's true, the gravitational pull from the center of the universe will cause the outward acceleration of the expelled matter to slow, stop, and then eventually reverse, pulling the matter back like a yo-yo.

As the matter gathering at the center of the universe gets more massive, gravity will collapse it into a black hole, or singularity. Eventually the matter will heat up enough to trigger another Big Bang, and we will have an Oscillating Universe in which this cycle is repeated throughout infinity.

In his story "The Last Question" (1956), Isaac Asimov writes an intriguing version of the Oscillating Theory in which a powerful super-computer, built by humankind, continues its existence in a dead universe long after the stars burn out. Finally, the God-like supercomputer figures out how to create a new universe out of the ashes, and begins with the words "Let there be light...."

Physicists at the University of Chicago say that the Big Bang was not a one-time event, but that Big Bangs occasionally happen at different points in space and time. Their calculations show that another Big Bang is likely to occur, although the probability of it taking place at any *particular* time or location is exceedingly small: 1 divided by 10 to the power of 10^{56}. Although this number is small, it is not zero, and since the universe is infinite, the next Big Bang will happen.

The latest theories (see the discussion of M-theory in "Alternate Universes") speculate that the Big Bang was not necessarily the creation of our universe—the beginning of all space, matter, and time—but rather the transition between our universe's current phase of existence and a previous phase.

After all, even if we accept the Big Bang theory, where did that first ball of super-heated, super-dense matter—the explosion of which produced everything now in creation—come from? What was there before it? Did this matter simply spring into existence? Did it arrive here from a parallel universe?

Cosmologists today don't doubt that the Big Bang took place, but now believe the universe existed in some form—both in the four dimensions we know and possibly others as well—before the explosion.

So the Big Bang probably created *our* universe, but not necessarily *the* universe.

Big Brother

OT ALL SCIENCE FICTION revolves around technological advances; sometimes the themes are sociological.

A prime example is George Orwell's classic 1949 novel *1984*, in which the government, known as "Big Brother," watches the activities of most of the population most of the time, ensuring conformity with approved government behavior. Big Brother is the figurehead ruler of the totalitarian government. His image is constantly visible on TV screens covering virtually all walls. The expression "Big Brother is watching" comes from the fact that these are two-way TV screens, allowing Big Brother and the government to watch everyone, even as everyone watches Big Brother.

Big Brother's rules are enforced—often through torture and brainwashing—by a paramilitary law enforcement squad known as the Thought Police. The primary goals are to instill conformity and obedience to the state in the citizenry, who are in turn monitored through the power of communications technology:

> In the past no government had the power to keep its citizens under constant surveillance. The invention of print, however, made it easier to manipulate public opinion, and the film and radio carried the process further. With the development of television, and the technical advance which made it possible to receive and transmit simultaneously on the same instrument, private life came to an end.
>
> Every citizen, or at least every citizen important enough to be worth watching, was kept for twenty-four hours a day under the eyes of the police and in sound of official propaganda, with all other channels of communication closed. The possibility of enforcing not only complete obedience to the will of the State, but complete uniformity of opinion on all subjects, now existed for the first time.

Three years before Orwell published *1984*, Ayn Rand came out with a short novel, *Anthem*, in which thought is controlled and regulated, and citizens are constantly watched by a Big Brother-like entity called the World Council. People are given numbers rather than names to take away their individuality.

The people of Rand's *Anthem* have no free choice; all their activities are dictated by the state. Even their lifetime vocations are chosen for them based on aptitude:

> Dare not choose in your minds the work you would like to do when you leave the Home of the Students. You shall do that which the Council of Vocations shall prescribe for you. For the Council of Vocations knows in its great wisdom where you are needed by your brother men, better than you can know it in your unworthy little minds. And if you are not needed by your brother men, there is no reason for you to burden the earth with your bodies.

Although it doesn't reference Big Brother by name, Ray Bradbury's novel *Fahrenheit 451* (1953) also portrays a society in which the government attempts to control and achieve conformity in individual thought by forbidding the reading of books, which are confiscated and burned by government-employed "firemen." The title, *Fahrenheit 451*, is a reference to the temperature at which paper burns.

While the notion of a benign image cloaking political surveillance may have originated with Orwell and reflected his apprehension about the cult of personality in the Soviet Union and elsewhere, credit for the concept of a world in which the citizens were under surveillance—or at least were capable of being located anywhere at any time—should go to H. G. Wells. Wells speculated about a future in which a central headquarters would keep track of everybody, and in which everybody would be able to connect with everybody else, which he considered progress.

Wells' vision is made feasible through technology in Arthur C. Clarke's and Stephen Baxter's *The Light of Other Days* (2000). By harnessing a wormhole, scientists create the WormCam, a device that enables the user to see anyone in the world, anywhere, at any time, thereby completely destroying individual privacy:

> A WormCam essentially enables us to locate a remote camera (in technical terms a "viewpoint") anywhere, without the need for physical intervention. WormCam intelligence—"wormint," as the insiders are

already calling it—is proving so valuable that WormCam posts have been set up to monitor most of the world's political leaders, friendly and otherwise, the leaders of sundry religious and fanatic groups, many of the world's larger corporations, and so on.

WormCam technology is intimate and personal. We can watch an opponent in the most private of acts, if necessary. The potential for exposure of illicit activities, even blackmail if we choose, is obvious. But more important is the picture we are now able to build up of an enemy's intentions. The WormCam gives us information on an opponent's contacts—for instance weapons suppliers—and we can assess knowledge factors like his religious views, culture, level of education and training, his sources of information, the media outlets he uses.

Thanks to advances in technology in general and the Internet in particular, the notion of a Big Brother watching us electronically—in ways that threaten to invade our privacy and erode our personal freedom—is gradually morphing from science fiction to science fact.

For instance, in the 1960s, the CIA took countless aerial photos of Soviet territory and a wide range of locations in other nations. This vast library of satellite survey imagery was, as one would expect, strictly top secret. The subjects ranged from nuclear plants in India to the backyard barbecues of American citizens.

Do these citizens have privacy? Not at all—now any individual can buy these satellite images from the CIA, according to a 2003 article in the *New York Times*.

The Freedom of Information Act, electronic commerce, and the Internet make a wide range of your personal information—from your credit records to your medical history—available to others, some of whom are permitted to look at your files and others who access it illegally.

U.S. intelligence agencies plan to launch nearly a dozen new satellites for ground surveillance within the next few years. They are also making arrangements to buy additional imagery from private companies. Given enough commercial and spy satellites, supplemented by aircraft, it may someday be possible to achieve constant, real-time surveillance of the entire planet—Big Brother coming to life!

The Patriot Act of 2001 allows the government to monitor citizens' transactions with travel agencies, car dealerships, hotels, real estate brokers, insurance agents, lawyers, and the post office. And they don't need a court order to access these records.

Another invasion of personal privacy comes from identity theft, achieved mainly by stealing and using other people's credit cards, either the actual card or more likely the card number. You don't even have to steal to gain access to an individual's financial information. U.S. Bancorp was accused of selling customer data to MemberWorks, a telemarketing firm, for $4 million and a twenty-two percent commission on sales made to those customers. Another company, CheckPoint, sold data files on 145,000 consumers to clients who turned out to be scam artists.

Even if you arrange to have your name taken off direct mail and telemarketing lists, governing bodies like the Direct Marketing Association cannot guarantee that marketers will leave you alone.

New legislation, the CAN SPAM act, has been enacted to cut down on "spamming"—unsolicited e-mails. But despite CAN SPAM and e-mail filtering software, spammers are still thriving—and many know a lot about you. Direct marketers maintain vast databases showing who you are, your age, where you live, your family status, your job, your income, even your product preferences. They freely share this data with anyone who will pay the ten cents a name or so they charge to rent their customer list to other marketers.

Our DNA also makes it more difficult for us to retain our privacy. For example, in 2005 an unruly London teenager spat into the face of a bus driver with whom he had a dispute. Police analyzed the spit and found a DNA match, enabling them to determine his identity.

New medical tests enable doctors to scan your genes to determine whether you are predisposed toward a particular disease, such as breast or prostate cancer. But if insurance companies can gain access to your genetic records, might they deny you health insurance coverage today based on the probability of your getting ill in the future? Someday employers may refuse to hire you based on your genetic predisposition to an illness you've never had and may never get.

Then there's the storing of your personal information in a variety of different computer databases, from business computers to government employees. A computer "hacker" can get past the security codes on these databases to download, read, and even alter your records on a whim. The number of computer/network security incidents reported to the Carnegie Mellon Software Engineering Institute rose from just six in 1998 to more than fifty-two thousand in 2002.

Soon, even your emotions won't be private. Researchers at the University of Montreal have created a working "mind probe" machine. Using brain imaging, the device can tell which of your emotions are aroused when you view a particular image. "Brain imaging has already delved into our personal lives," writes Helen Phillips in *NewScientist* (July 31, 2004), "Among other things, it has been used to investigate love, personality traits, political leanings, racial prejudice, tendency to violence, deception, and moral reasoning."

In the 2002 motion picture *Minority Report* (based on a 1956 Philip K. Dick story with the same title), "precogs" (people with precognitive powers) are used to scan the populace for criminal thoughts. When they detect a person who is going to murder someone in the future, the data is transmitted to a screen, alerting the police to go out and arrest the suspect before he commits the crime, an ethical dilemma in and of itself. Studies are underway to test whether brain imaging could predict who is likely to commit a crime, and whether people are lying or have false memories.

MIT scientists have created software that allows computers to analyze voice mail messages you leave and accurately indicate your emotional state when you called: happy, sad, excited, calm, urgent, not urgent, formal, and informal.

Big Brother may also be able to track you down by the cell phone you carry. Michigan Technological University is patenting a vehicle-mounted transmitter that causes all nearby cell phones to respond with their identity numbers. If you are hiding out in a house, the vehicle will be driven around the neighborhood in which Big Brother suspects you are hiding. When your phone is hit with the signal, it sends your ID number, which is detected by a receiver, mounted on the vehicle. The device then sends out a clocking signal. By measuring the time it takes to get a response, it can calculate how far away you are, pinpointing your position. Then Big Brother can come and take you away.

The U.S. Food and Drug Administration (FDA) has approved the use of implantable radio frequency identification (RFID) tags to identify patients and retrieve their medical records. The U.S. Department of Homeland Security has ordered a million RFID tags to be implanted in the passports of American citizens. Customs officials will be able to scan passports with an electronic reading device, automatically bringing up personal details and biometric data on the passport holder.

The "Big Brother" aspects of implantable RFID tags are frightening. Doctors could refuse to treat a patient, or insurance companies to cover them, based on the health history in their RFID tag. Police and security officers with handheld scanners, which can read RFID data at a distance, could pick people out of a crowd based on nationality, religion, or other retrievable data.

In Japan and Denmark, school authorities are implanting RFID chips in kids' schoolbags, name tags, and clothing. Scanners at school gates and other key locations can identify each child by his or her RFID chip, allowing schools to track students' movements. In Mexico, the attorney general and 160 members of his staff have had RFID chips implanted in their arms for security purposes. And Delta Airlines recently announced it will be using RFID to track passengers' luggage.

Researchers in the U.S. are developing software capable of identifying people by the TV shows they watch. In one trial test, the "TV ID" system achieved an eighty-two percent accuracy. Based on this technology, someone monitoring TV viewing in your house could determine who was home based on the programs being watched.

Technology is even enabling old-fashioned fingerprinting to become a more accurate means of tracking and identifying people. According to a study by the National Institute of Standards and Technology, computerized systems that automatically match fingerprints to suspects' records are accurate more than ninety-nine percent of the time.

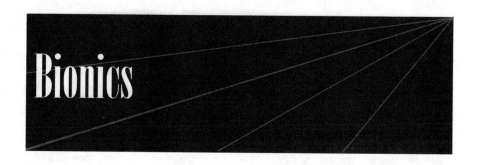

Bionics

STEVE AUSTIN. A MAN BARELY ALIVE. Gentlemen, we can rebuild him. We have the technology. We have the capability to build the world's first bionic man. Steve Austin will be that man. We can make him better than he was before. Better. Stronger. Faster.

Science fiction fans thrilled to hear these words, the opening of the 1970s television series *The Six Million Dollar Man*, about the world's first bionic man.

Based on the 1972 novel *Cyborg* by Martin Caidin, the TV show featured the super-heroic exploits of Steve Austin, a bionic man given super strength, speed, and sight through the technological miracle of a bionic eye, bionic arm, and two bionic legs.

In the novel, Dr. Rudy Wells, one of the scientists who give Austin his bionics, explains why the bionic limbs are so much more powerful than human arms and legs:

> The energy for articulation—the electrical impulse sent along the nerve network by brain command—isn't great enough to cause the bionics limb to react through muscle contraction and movement. The voltage is far too low, and we can't avoid the fact that we're working with different systems. Different materials, in fact. The body provides electrochemical reaction throughout the entire system. That's impossible with the bionics elements.
>
> So we compensate for this. What the electrical impulse originating from the brain must do, then, is to trigger another energy source within the bionics limb. In this case we provide additional energy. The best way to do this is not through solenoids, which could result in a staccato or jerking movement, but through the latest advances in electrical motors.

Steve Austin had a bionic right arm; in the Star Wars film saga, both Luke Skywalker and his father Anakin lose a hand in light saber duels

(Luke's is cut off by Darth Vader) and have them replaced with bionic hands. In the 2004 film *I, Robot*, Will Smith plays a police officer whose bionic left arm allows him to hold his own in hand-to-hand combat with rogue robots that have evolved to the point where they can violate the Three Laws of Robotics (see "Robots").

What, exactly, are bionics? *Dorland's Illustrated Medical Dictionary* defines bionics as "the science concerned with study of the functions, characteristics, and phenomena found in the living world and application of the knowledge gained to new devices and techniques in the world of mechanics." Closely related to bionics is cybernetics, which *Dorland's* defines as "the science of the processes of communication and control in the animal and in the machine." In science fiction, bionics refers to high-tech artificial limbs, organs, and other body parts used to replace injured or damaged tissue, bone, and muscle in humans. In *Cyborg* Caidin gives this definition of bionics: "biology applied to electronic engineering systems."

Steve Austin was not the first science fictional character to receive artificial parts. In the 1939 movie *The Man They Could Not Hang*, Boris Karloff plays Dr. Henryk Savaard, inventor of a mechanical heart that keeps him alive after he is hanged for murder. Greg Bear's *Queen of Angels* (1990) includes the use of cybernetics, as does Frank Herbert's *Destination Void* (1966).

But now science fiction, and its science predictions, is becoming science fact—and the first bionic men and women are being built even as you read this chapter. One out of ten Americans—twenty-eight million people—already have synthetic body parts.

The key to creating bionics is to get computers to respond to human nerve impulses. If you can do that, you can use "embedded microprocessors"—small computers built into bionic organs and limbs—to control the bionic part.

Here's how it works: An electrode is implanted into the patient's brain. Specifically in the motor cortex, which controls body movement. When the patient thinks about moving, his brain generates a signal. Normally, this signal would be transmitted through nerves, which in turn would activate the muscle in the part of the body the person wants to move. In bionics, the signal generated by the brain is picked up by the electrode, which amplifies and transmits the signal to the embedded microprocessor. The microprocessor then controls the motor or other motion-producing component in the bionic body part.

Bionics for virtually every part of the human anatomy are under development today:

- In *The Bionic Woman*, a spin-off of *The Six Million Dollar Man*, Lindsay Wagner was able to hear with a bionic ear. Now, bionic ears have become a reality. Known as "cochlear implants," today's bionic ears are high-tech implants that allow deaf people with damaged hair cells in the inner ear to hear. Parts of the bionic ear, the receptor and transmitter, rests behind the real ear, just like some hearing aids. The other two components, a receiver and an electrode, are surgically implanted in the patient's ear. The receptor picks up sound waves from the air, and converts them into digitized signals. The transmitter sends these signals to the receiver, which converts them to electric impulses. Finally, the electrode transmits these impulses to the brain, where they are interpreted as sound.

- Remember Steve Austin's bionic eye? A Canadian man who lost his eyesight in a snowmobile accident paid surgeons $115,000 to drill a hole in his skull and implant an array of electrodes onto the surface of his brain. Only a handful of people have received the procedure so far. Success is limited, but some vision has been restored: the Canadian patient sees well enough to avoid obstacles when driving a car in a parking lot. How does the bionic eye work? Surgeons implant a tiny silicon chip, containing five thousand solar cells, in the patient's retina. The solar cells, each smaller than the head of a pin, are connected to an electrode, which in turn is in contact with neural cells that normally receive signals from the retina. When light strikes the chip, the solar cells send a visual signal to the neural cells. The brain interprets the signals as images, just as it would with the patient's original retina.

- Researchers at MIT have developed "artificial muscles" composed of electronically conductive polymers. When electric signals are sent to the polymers, they contract, simulating the motion of a natural muscle—but with one hundred times the strength. An article in *American Demographics* (August 2002) observes that these polymers "can be formed into artificial muscle to give ordinary folk superior strength." One application in the works: a bionic sphincter that can restore urinary control to men whom prostate surgery has rendered incontinent.

- Another research group has developed a bionic ligament. Made of a modified silk, the bionic ligament is used to replace torn ligaments in knee joints. The modified material allows the body's natural cells to infiltrate and attach to the silk. Natural ligament begins to grow around the silk fibers. Eventually, the silk dissolves. At that point, the patient has grown enough natural ligaments to support the knee without prosthetics.

- Harvard physicist Robert Westervelt attaches micron-size magnetic beads to individual cells, which float in fluid above a grid of computer-controlled magnetic coils. The cells are moved magnetically by turning the cells on and off from a control console. The objective: assemble the cells into an artificial (bionic) skin.

- Amputee Mark Marich walks with the aid of a bionic leg. His high-tech artificial limb contains sensors in its ankle and knee that relay force and position to its on-board computer fifty times a second. Based on feedback from the sensors, the computer activates motors that hydraulically adjust the bending of the knee for smooth walking. Although Marich's bionic leg is removable, it doesn't have to be. Swedish doctors have developed a way to permanently attach bionic arms and legs to the patient. Known as osseointegration, the technique calls for doctors to drill holes into the bone. Titanium anchors are dropped into the holes. The bone grows around the titanium, holding it solidly in place. The bionic limbs are then clamped to the titanium anchors.

- The "myoelectric" technology to make prosthetic arms has been around for decades. The arm straps on to the amputated limb. When the user contracts his muscles, sensors in the prosthetic detect the electrical impulses and open and close the hand. A biomedical engineer at Rutgers University has designed a more advanced hand. He built an artificial hand powered by motors from model airplane kits. The motors were attached to the fingers using fishing line as the "tendons." Other engineers are building bionic hands without motors. They use bionic "muscles" made of plastic bonded to metal. The metal contracts when an electrical signal is applied, just like an organic muscle contracts when a nerve sends a signal.

- As the MIT research on bionic muscles demonstrates, bionics isn't limited to large body parts, such as eyes, ears, arms, hands, and legs. Dr. Gerald Loeb, a biomedical engineer, has developed a bionic nerve, called a BION. When patients have a stroke or spinal cord injury, their nerves don't send signals to the muscles, and the muscles atrophy. The BION solves the problem by mimicking nerve impulses from the brain. Injected into the patient's body with a large needle, the BION is placed where nerve endings and muscles meet. When the muscles are ready to be exercised, a magnetic transmitter coil is placed over the spot on the body where the BION is implanted. The coil sends power and command signals to the BION, prompting it to emit electrical impulses that activate the muscle.

- British engineers are experimenting with a "bionic tooth"—an artificial tooth containing an implanted microprocessor. Their eventual goal is to weave computer circuitry directly into a person's living skin. The augmented skin would enable the person to directly access computer data and images.

- Biochemists have developed "bionic blood," an artificial blood substitute. Unlike real blood, there's no need for blood-type matching, so the bionic blood is a match for every patient. The molecules of bionic blood are up to a thousand times smaller than real red blood cells, so they can squeeze past obstructions, such as a blood clot that is preventing a stroke victim from getting oxygen delivered to his brain. The small size of the bionic blood molecules could conceivably make getting a transfusion of the artificial blood a treatment for hypertension. The smaller and more pliant the blood cells, the less likely they are to cause pressure build-up in the circulatory system.

- In 1969, Dr. Denton Cooley implanted an artificial heart into Haskell Karp, a patient dying of cardiac disease. The artificial heart kept Karp alive for sixty-five hours, until he could be given a heart transplant. In 1982, Dr. William DeVries implanted an artificial heart, the Jarvik-7, into Barney Clark, a Seattle man dying of heart failure. Unlike Karp's heart, the Jarvik-7 was intended to be permanent. It kept him alive 112 days. Artificial heart experiments have been conducted for decades, but the technology is advancing to the point where bionic hearts could soon be an alternative to donor hearts. Made of titanium and plastic, the self-containing bionic heart can be implanted permanently into a patient, with no external power or pump connections. One patient who received a bionic heart made by Abiomed was doing well more than seven months after his operation. In October 2004, the FDA approved, for the first time, a "totally artificial heart" (TAH). The TAH is designed to replace a patient's diseased heart until a donor becomes available.

- The newest goal for artificial organ development is an artificial or "bionic" liver. The liver is the largest organ in the human body and performs more than five hundred functions. One of the most significant is the liver's ability to detoxify the body and cleanse the blood. Without a healthy, functioning liver, you could survive for only a short period of time. The first bionic livers are not designed to be implanted in your body. Rather, the patient is hooked up to an external artificial liver, in much the same way that dialysis performs the function of failing kidneys. In the artificial liver, a bioreactor—a small vessel capable of sustaining living cells—is equipped with membranes containing colonies of living liver cells. As the patient's blood passes through the bioreactor, the liver cells perform the functions normally carried out by the patient's own liver. The goal is to extend the patient's life until a donor liver can be found for transplant or the patient's own liver heals (the liver is the only organ in the body that can fully regenerate). Future designs may incorporate a much smaller bioreactor that can be implanted into the patient's body as a longer-term liver replacement for patients for whom donor livers are not available.

- A bionic spine offers new hope for patients with spinal injuries. In a laboratory experiment, injecting the polymer polyethylene glycol (PEG) into dogs within three days of a spinal injury helped mend damaged nerves and improved their chance for a recovery. Of nineteen dogs treated for paralysis caused when their spines ruptured, sixty-eight percent regained the ability to walk. Injecting the polymer helps patch holes in damaged cell membranes and fuses the membranes together. In an earlier experiment, PEG was shown to partially restore connections in guinea pigs whose spines were completely severed.

- Nanotechnology is helping researchers at Purdue University create stronger artificial joints. The artificial joints provide a polymer matrix holding single-wall carbon nanotubes. Osteoblasts, the body's bone-forming cells, attach themselves to the nanotubes, which possess surface bumps as wide as one hundred nanometers. The nanometer-scale bumps mimic the surfaces of proteins and natural tissues, enabling the osteoblasts to adhere better to the artificial joint, promoting the growth of new bone cells.

- A new bionic implant called the Orgasmatron (I kid you not) can give women orgasms on demand, according to its inventor, Dr. Stuart Meloy. The device is a small spinal cord stimulator, about the size of a pacemaker, embedded in the patient's lower back and activated by a handheld remote control.

Black Holes

$\mathbb{B}$ LACK HOLES HAVE PLAYED a part in countless science fiction stories. The black hole is a particular favorite of Frederik Pohl, whom I met in the late 1970s. He has written a number of sf novels dealing with black holes, most notably *Beyond the Blue Events Horizon* (1980) and *Gateway* (1977).

Scientist John Wheeler introduced the term "black hole" to describe a solar gravitational collapse in 1969, although it was first pointed out in the eighteenth century by John Mitchell and the Marquis de Laplace. Stories that picked up on the idea of black holes and related relativistic phenomenon include:

- Poul Anderson's "Kyrie" (1968).
- Robert Silverberg's "To the Dark Star" (1968).
- Brian Aldiss' "The Dark Soul of Night" (1969).
- Larry Niven's "The Hole Man" (1974).
- Barry N. Malzberg's *Galaxies* (1975).
- Arthur C. Clarke's *Imperial Earth* (1975).
- Frederik Pohl's *Gateway* (1977).
- John Varley's "Lollipop and the Tar Baby" (1977).
- Ian Wallace's *Heller's Leap* (1979).
- David Brin's *Earth* (1990).
- Paul J. McAuley's *Eternal Light* (1991).
- Roger MacBride Allen's *The Ring of Charon* (1990).

A black hole is an extremely dense star. The star has collapsed, increasing its density to the point where nothing, not even light, can move fast enough to escape its tremendous gravitational pull. Since nothing can escape, the star is like a "bottomless pit" or hole in space, sucking in everything that comes near. It is black because no light shines from it. Hence the name "black hole."

Physicists first theorized the existence of black holes, but science fiction writers brought these fascinating objects to the public awareness. In his 1974 novel *The Forever War*, Joe Haldeman wrote about an interstellar war using black holes for travel between battles. Black holes are also used as stargates in Joan Vinge's *The Snow Queen* (1980).

Our Sun will probably never become a black hole—it simply isn't massive enough. Stars weighing 1.4 solar masses or more end up as black holes. Let's take a hypothetical star that weighs between 1.4 and 10 solar masses, and see how it becomes a black hole.

The star starts out as a ball of hydrogen. Hydrogen, the lightest atom of all the elements, consists of a single electron orbiting a proton. Two hydrogen atoms bonded together form a molecule of hydrogen gas. The hydrogen begins to undergo nuclear fusion at the gas ball's center when the temperature there reaches about four million degrees Celsius. The star's energy comes from this fusion of four hydrogen nuclei into one helium nucleus. In the process, a small amount of mass is converted into energy.

In the case of our own Sun, four million tons of its mass are converted into energy each second. Einstein's well-known mass-energy equation $E = mc^2$ can be used to show that our Sun puts out 4.4 septillion horsepower per second. This energy production tends to make the star expand. But the expansion force is counter-balanced by a contraction force. The contraction force is due to the weight of the star's outer layers. The outer mass tends to collapse inward toward the star's gravitational center.

When all of the star's hydrogen is used up near the core, and the core is mainly composed of helium, the balance of forces is upset. The gravitational forces collapse the core because the expansion force is no longer present as the nuclear fusion of hydrogen into helium has ceased.

The fusion reactions still continue in the layer around the center, and the outermost layers expand. The gravitational collapse of the core causes temperatures there to rise until it is hot enough to initiate helium fusion. Helium, not hydrogen, then becomes the star's fuel. The helium fusion produces even heavier elements. The star has a very hot core, but has expanded tremendously in the outer regions. Stars of this type are called red giants because of their expanded size and red color.

Let's go back to our hypothetical star. For a star of the mass range we have chosen, the nuclear reactions in the core go out of control as that region collapses. An explosion occurs, and the star blows most of its mass out into space in what is known as a supernova. What remains of the star's matter shrinks down into a small, dense body. Protons collide with electrons to form neutrons, leaving us with a super-dense star whose core is composed of solid neutrons—a neutron star. These neutron stars (see the chapter on "Neutron Stars" for a more complete description) have densities as high as ten million tons per cubic centimeter—about nine trillion times denser than water.

If the star that we are talking about weighs between 1.4 and two solar masses, it ends its life as a neutron star. But if its mass is greater than two solar masses, further contraction occurs. The neutron star becomes what is known as a collapsar—an object collapsing toward its center under gravitational attraction. Our hypothetical star is now a collapsar. It is shrinking, and its radius decreases until it reaches a certain critical value called the Schwarzschild Radius. When the radius reaches this value, the spherical mass becomes so dense that even light cannot escape it. Now the gravitational field of the star is tremendous, and an event takes place that no astronomer will probably ever witness.

Suddenly, all of the star's matter is compressed to the center point. This cannot be seen because it happens too quickly to observe—the collapse takes place in a time interval on the order of millionths of a second, emitting a powerful burst of gamma rays. The density and gravitational force of the newly formed black hole are infinite at this point—which is called a singularity—and the matter of the star has literally become crushed out of existence.

Surrounding the singularity is a volume of space into which matter has fallen and nothing can escape—a black hole. The radius of the black hole measured from the center is the Schwarzschild Radius.

If our hypothetical star had been heavier than ten solar masses, we also would have ended up with a black hole. With a star that heavy, once the core has used up its hydrogen, nothing can stop its collapse. No supernova outburst need occur; its own mass will cause it to rapidly collapse into a black hole.

Stars weighing 1.4 solar masses or less become white dwarf stars when their nuclear fuel is exhausted. The red giant collapses until it is a white dwarf, and stabilizes at that point (no further collapse taking place). While not as compact as neutron stars, white dwarf stars are still quite dense; one thimbleful of white dwarf material may weigh over a ton. White dwarfs owe their density to the fact that they are composed of what is known as degenerate matter.

In ordinary gaseous matter, the space between atoms is so much greater than the diameter of the atoms themselves that the atoms may be considered point particles for all practical purposes. They fly about at random, and the pressure exerted by the gas is due to these point particles bouncing off other particles. In degenerate gaseous matter, the atoms are packed together so closely that the distance between the atoms is not much greater than the diameter of the atoms themselves. In this case the size of the atoms is significant and they can no longer be considered point particles. Now, electrons orbiting the nuclei make up most of the volume of the atoms. The electrons resist being squeezed together in such close quarters. This resistance is called degeneracy pressure, and is the pressure the gas exerts against anything trying to compress it. This pressure holds the white dwarf up.

However, just as there was for a neutron star, there is a certain critical mass for a white dwarf beyond which it cannot resist further collapse. This is known as the Chandrasekhar Limit, and is 1.4 solar masses. If a white dwarf should incorporate more mass into its structure and exceed 1.4 solar masses in weight, it would become a collapsar and end up a black hole.

In the region of a black hole, the laws of physics behave contrary to ordinary experience. In particular, the laws of relativity show pronounced effects in these regions. Gravity, for instance, is so intense that even the speed of light is not sufficient to escape it.

The boundary surrounding the black hole region of space is called the event horizon. Frederik Pohl referenced this in the title of his 1980 novel, *Beyond the Blue Event Horizon*, which deals with the Gateway— a portal that was constructed and abandoned by an unknown species. The Gateway contains hundreds of modules which transport voyagers to predetermined locations throughout the universe.

Outside the event horizon an observer cannot have any knowledge of what is going on inside the event horizon, i.e., we cannot see what is going on inside a black hole if we view it from the outside. Reason: The gravity of the black hole is so intense that even light cannot escape the event horizon boundary.

The event horizon is a barrier to communication of any kind, so an observer who has fallen inside a black hole cannot report to those outside what he is experiencing. He can't ever get back outside the event horizon to report what he's seen, since nothing can escape from a black hole. Thus, it seems that we can never directly observe what happens inside a black hole.

Suppose we decide to send an astronaut inside a black hole, anyway. The astronomer will observe him from a safe distance away through a telescope. The first thing the astronomer observes is a very pronounced red shift in the vicinity of the black hole. (What is a red shift? We noted earlier that the tremendous gravitational field of a black hole is so strong that not even light can escape from it. All gravitational fields have an effect on light, which loses some of its energy whenever it has to struggle against gravity. We know that energy is proportional to the frequency of the light wave, so when the light loses energy, its frequency decreases. Since frequency and wavelength are inversely proportional, this decrease in frequency causes an increase in wavelength. This "shift" toward an increase in wavelength is called a red shift, because red light has a longer wavelength than any other visible color. The red shift, a quantity denoted by the letter z, is defined as the shift in wavelength per wavelength of light emitted. For example, consider a beam of light with a wavelength of 5,000 angstroms. If the light is shifted 250 angstroms, z = 250/5000 = 0.05. At the center of a black hole gravity is infinite, and so the light trying to escape loses an infinite amount of energy. Hence the red shift becomes infinite in the region of the black hole.)

Let us consider the astronaut falling toward the black hole as observed by the astronomer from a distance. Each is wearing an identical stopwatch. As the astronaut approaches the event horizon, time itself actually slows down for him. This is to be expected, since Einstein's theory of relativity predicts that time slows down in the region of an intense gravitational field.

For each second that the astronaut's stopwatch ticks off, the astronomer's watch ticks off 1 + z seconds. Since the red shift z increases as one approaches the event horizon and eventually becomes infinite, the quantity 1 + z also becomes infinite. This means that as one second passes for the astronaut (who is right at the event horizon), an infinite amount of time passes in the universe. In other words, time is frozen at the event horizon. The astronomer peering through his telescope would see the astronaut falling slower and slower toward the event horizon, until he seemed frozen forever at the event horizon.

Einstein showed us that time, mass, and length are relative, depending on your point of view, or your reference frame, as it's called. For example, to a man standing still, a rocket ship weighing tens tons at rest would weigh almost twenty-three tons if traveling at ninety percent the speed of light. However, if the man was flying alongside the rocket ship and weighed it, he would find it weighed ten tons. If he is in the same reference frame as the rocket, it doesn't gain any mass. In a different reference frame, it does. By the same reasoning, the astronomer and astronaut are in different reference frames, this time the difference being due to gravitation and not velocity. To the astronomer outside the black hole's pull, it appears as if the astronaut is frozen at the event horizon. But what is it like in the astronaut's frame of reference?

As he approaches the event horizon, time runs perfectly normally from his point of view. As he falls, he sees nothing in front of him; a black hole looks like just that—a hole. At this point he might be killed or at least seriously injured by something that affects oceans on Earth, namely, the tide. A tidal force is exerted by a mass on a body. We know that the gravitational force decreases in strength with distance. It follows that the gravitational attraction on a body is strongest at that point on the body nearest the mass, and weakest at that point on the body farthest away from the mass.

The Moon's gravity acting on the Earth causes the oceans to bulge toward the Moon (at the point nearest the Moon) and away from the Moon (at the point farthest away). This gives us two tidal humps on opposite sides of the Earth, as if the gravitational forces were "stretching" the waters.

Now we are well aware of the fact that a black hole exerts a great deal more gravity on the astronaut than the Moon exerts on the Earth. For a black hole of ten solar masses, the tidal force at the event horizon is about a hundred million million times greater than the tidal force one would feel standing on the Earth. The tremendous tidal force does to the astronaut something similar to what happens with the ocean.

If the astronaut approaches the black hole feet first, the gravitational pull on his feet will be considerably stronger than that on his head. So strong is the force that the astronaut will be stretched out into a long cylinder. He will be torn apart on a sort of cosmic torture rack. If the astronaut isn't killed by tides at the event horizon, the infinite gravitational force at the black hole's center will surely do the trick. Larry Niven's Hugo-winning story, "Neutron Star" (1966), is about the tidal effect of a neutron star on an astronaut (who, by the way, survives).

But whether he does it dead or alive, the astronaut will pass through the event horizon. Once inside the black hole, he is pulled toward the center point where all the mass of the hole is contained—the singularity. Since matter is crushed out of existence at the singularity, all objects lose their identity. Our astronaut is no longer an astronaut; he is merely mass that has been added to the black hole. He no longer exists in our universe. But it doesn't really matter, since once past the event horizon he was out of the picture forever.

The actual observation of a black hole in space would strongly confirm our present theories of relativistic physics. But how can we possibly detect a black hole far out in space? It is invisible, emitting no radiation whatsoever. Unless we are close enough to feel its gravity, it seems, we won't know if it's there or not.

Stars and star-like objects exist not only as single objects, but also in pairs. A system of two stars orbiting one another is called a binary star. It turns out that if one of the stars is a black hole, we should be able to detect it through its interaction with the other star.

An X-ray binary is a system in which a normal star is paired with a super-dense star. The dense star may be a white dwarf, a neutron star, or a black hole. Matter streams from the normal star and follows a spiral path until it falls into the denser star. As the matter falls into the dense star energy is gained and converted into X-rays. This is because the matter is ionized plasma, and has an electrical charge. The X-rays are emitted into space (hence the name "X-ray" binary) and can be detected by the astronomer's instruments in orbit above the Earth.

It is assumed that the super-dense star has a magnetic field. For a neutron star this field is not necessarily aligned with the axis of rotation of the star. A black hole possesses axial symmetry, meaning its magnetic field is aligned with its rotational axis.

The X-ray signal will be a regular pulse if the magnetic field is skewed, so regular pulses indicate a neutron star. If the pulse is sharp and irregular, the magnetic field is aligned with the axis of rotation and we have a black hole. Each burst of X-rays is caused by matter spiraling toward the super-dense companion star, and the bursts get narrower as the spiral orbit gets smaller.

Using this information, astronomers scan the skies for black holes. Remo J. Ruffini, a Princeton astronomer, found the first black hole, Cygnus X-1, in 1971. Since then, many other black holes have been observed.

One giant black hole in the Perseus Cluster, a group of galaxies located 250 million light years from Earth, has been emitting a musical note—a B-flat about fifty-seven octaves below the middle C in a standard piano keyboard—for billions of years.

Another giant black hole, in a galaxy about seven hundred million light years away, completely ripped apart a star the size of our Sun within days, swallowing the equivalent of one Earth every ten minutes. The star got caught in the black hole's powerful gravitation field when its trajectory brought the two objects close together. The black hole, as massive as one hundred million suns, stretched the star until it pulled apart.

Our own galaxy, the Milky Way, also contains a giant black hole between three to four million times the size of our Sun. Fortunately, this black hole is located in the center of the Milky Way, about twenty-five thousand light years away from Earth. So it won't pose a danger to us any time soon.

At the other end of the size spectrum from giant black holes are miniature black holes: some physicists today believe that when cosmic rays, the most energetic particles in the universe, strike the Earth's atmospheres, some of these collisions could form miniature black holes.

These miniature black holes would have a diameter of just 1/100 of a quadrillionth of an inch and weight equal to a thousand protons. But they evaporate in a fraction of a second because, unlike ordinary, more massive black holes, micro-black holes leak energy.

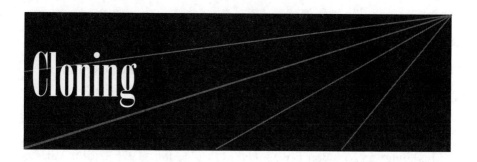

Cloning

LONING IS THE PROCESS of making a genetic twin or duplicate of a living organism. The nucleus from one of the organism's cells is implanted in an egg cell (from a different organism) whose nucleus has been removed. The combined cell is stimulated so it begins to divide, and is then implanted in an organic or artificial womb. The resulting organism would be an identical twin—biologically—to the nucleus donor.

Cloning has long fascinated science fiction writers and the general public. Often in science fiction movies, a man is cloned, creating an instant double. In reality, while the clone is a genetic duplicate of the original, the clone is younger. If I clone you, your clone will be a newborn, while you will be whatever age you are now.

Also, although you would be biological duplicates, your clone would have none of your memories. In short, he or she would be a completely different person.

The classic cloning novel in science fiction is Aldous Huxley's *Brave New World,* published in 1932. In the novel, children are no longer conceived through sexual intercourse but are grown through a combination of artificial insemination of eggs with collected sperm and cloning at "hatcheries." Once an egg is fertilized, it is duplicated through "Bokanovsky's Process," cloning ninety-six genetically identical individuals from the single fertilized egg.

Different batches of cloned embryos are given different nutrients as they develop to create workers with desired characteristics of strength, height, physical attractiveness, and intelligence—alphas, betas, gammas, deltas, and so on. Therefore a person's lot in life is set before he or she is even born.

Gilbert Gosseyn gets cloned, although the word isn't used, in A. E. van Vogt's *The World of Null-A* (1945). In Theodore Sturgeon's "When You Care, When You Love" (1962), an individual is cloned from his cancer cells. The Andrology Institute of America (AIA) claims it has created clones of dead people by fusing their cells with cow eggs stripped of their DNA. Cows have already been cloned from two-day-old pieces of steak, so it might be possible to clone people from corpses.

Ursula K. Le Guin's "Nine Lives" (1969) is a classic exploration of cloned relationships. Other sf novels with a cloning theme include Pamela Sargent's *Cloned Lives* (1976), Kate Wilhelm's *Where Late the Sweet Birds Sang* (1976), Ben Bova's *The Multiple Man* (1976), John Varley's *The Ophiuchi Hotline* (1977), and Fay Weldon's *The Cloning of Joanna May* (1989).

Poking a bit of fun at the whole idea of cloning, science fiction writers Isaac Asimov and Randall Garrett composed "The Clone Song." The first verse is as follows:

Oh, give me a clone
Of my own flesh and bone
With its Y chromosome changed to X
And after it's grown
Then my own little clone
Will be of the opposite sex....

Science is quickly catching up with science fiction as far as cloning is concerned. The first clone of a vertebrate was produced in 1967, when British biologist John Gurden successfully cloned a South African clawed frog. Gurden removed a cell from the intestine of an adult frog. He then transferred it to an egg cell in another frog of the same species. The ovum developed into a new frog that was a genetic replica of the frog that donated the intestinal cell.

In the 1970s, researchers discovered that they could clone single cells. They used this technique to produce "hybridoma" cells.

Hybridomas are essentially immortal cells that grow indefinitely when kept in a culture. The hybridomas secrete antibodies, which the body can use to fight infection and other disease.

Science writer David Rorvik created a sensation with his 1978 book *The Cloning of a Man*, which claimed that an eccentric millionaire had financed an experiment resulting in the successful creation of a clone of himself. Two years later, a British court declared the book to be a fraud and a hoax. The following year, Dr. Landrum Shettles published a paper in the *American Journal of Obstetrics and Gynecology* in which he reported creating three human embryos cloned from adult human cells.

In 1980, Martin Cline placed a gene from one mouse into another, and showed that the gene continued to function after the transfer. The following year, scientists at Ohio University placed rabbit genes into mouse eggs, proving that genes can be transferred into different species. The Ohio researchers then transplanted these mouse/rabbit hybrid eggs into the reproductive systems of female mice who gave birth to baby mice. When the baby mice with the rabbit genes grew up and mated, some of their offspring also contained the rabbit genes, proving that the new animal could also pass on the transplanted gene to future generations.

In another experiment that same year, Swiss scientists took a nucleus from a mouse embryo and transplanted it into a mouse egg from which the nucleus had been removed. The mouse born from this experiment was genetically identical to the embryo from which it had gained all of its genes.

More recently, a team of researchers at Tokyo University created baby mice from two eggs. The eggs were from genetically altered mice and could therefore produce IFG-II, a growth factor that ordinarily comes only from the male genome and is necessary for embryonic development. These male-like gametes were fused with normal mouse eggs, chemically activated, and implanted in surrogate mice which gave birth to a litter of baby mice free of apparent genetic defects.

In 1997, Scottish scientist Ian Wilmut and his team successfully cloned an adult mammal. They were the first to accomplish this feat. Wilmut used a small pipette to suck out the nucleus from an unfertilized sheep egg, leaving the egg's cytoplasm empty. The nucleus from another adult sheep—the subject being cloned—was placed into the empty egg. Once this was done, the egg contained a complete set of genes, just as if it had been fertilized the natural way, by a sperm. The difference: all the genetic material in the egg was from a single parent—the animal that donated the nucleus. Electric pulses were used to get the nucleus and egg to fuse and start dividing, creating an embryo. The embryo was placed in a third female sheep, which gave birth to Dolly, a genetically identical replica of the donor sheep.

Some scientists believe that cloning produces defective duplicates, resulting in unhealthy clones. In one study, scientists at the U.S. Department of Agriculture lab in Lubbock, Texas, found that cloned pigs had immune systems much weaker than non-cloned pigs in a control group. And Dolly the sheep's premature death at age six, caused by lung disease, only contributed to the worry that mammals created through cloning are flawed copies of the original, rather than perfect duplicates.

Dolly's cells showed signs of aging more typical of older animals. One theory is that the original genetic blueprint of the subject being cloned wears out. But in their book *The Science of Jurassic Park and the Lost World* (1997), Rob DeSalle and David Lindley give a different explanation of why cloning is doomed to produce imperfect doubles:

> One of the reasons we each possess two sets of chromosomes is that this arrangement guards against a variety of genetic errors. Broadly speaking, you have two copies of every gene so that if one of your chromosomes happens to contain a faulty copy, the other will contain a working one.
>
> It's been estimated that everyone has perhaps a dozen potentially fatal genetic errors in his or her genome; the reason they aren't in fact fatal is that there is a good copy of the relevant gene as well as the bad one. But that also means that if you were to take a single set of chromosomes and duplicate it, the faked-up double set may well contain lethal genetic instructions, simply because any flaw in one genome would be faithfully reproduced in the other.

Another explanation for imperfect clones may be in the cloning technique itself: experiments at the University of Pittsburgh showed that the reason many cloned monkeys had genetic abnormalities could be that the technique of removing the nucleus of an egg with a micropipette was destroying some of the molecules the cells need to divide property. Results improved when gentler techniques were used for nucleus removal.

Hundreds of animals have been cloned worldwide. Many appear healthy and perfectly normal. But some cloning experiments have produced deformed fetuses that died in the womb with oversized organs. Others were born dead or died shortly after birth, and were abnormally large.

Scientists at the Oregon Regional Primate Center in Oregon announced that they have cloned monkeys from embryonic cells. Researchers elsewhere have also used similar techniques to clone cows and rabbits. Yet another team of scientists recently created a cloned mule embryo in a test tube, and then used a horse as a surrogate mother to give birth. And Stanford University has announced the creation of a $12 million center that will clone embryos for research purposes.

Cloning is already being used to clone endangered species. In China, scientists are planning to try cloning to save the Yunnan snub-nosed monkey, a species on the verge of extinction.

Australian researchers, meanwhile, are making plans to clone the already extinct Tasmanian tiger. The last known specimen died in 1936, and the clone will be made from DNA extracted from a Tasmanian tiger pup preserved in alcohol in 1866.

In April 2003, Advanced Cell Technology announced that they had successfully cloned two baby bantengs using cells from a specimen that had been dead for several decades. The banteng is a wild bovine species from Southeast Asia.

The operation was decidedly high-tech: The clones were produced from the frozen cells of a male banteng who had died more than twenty years ago without producing offspring. The DNA from these cells was transferred into cells from ordinary domestic cows, which also gave birth to the banteng calves.

Scientist and sometime sf author Charles Pellegrino has speculated that extinct species might be recreated from DNA recovered from specimens founded embedded in amber; Michael Crichton turned the idea into *Jurassic Park* (1990), which suggests that extinct species (in the case of the book and movie, dinosaurs) can be recreated by cloning cells from preserved specimens. Now a team of French scientists proposes to make this a reality by cloning a twenty-three-ton, eleven-foot-tall male wooly mammoth preserved in the frozen tundra of Siberia for twenty thousand years. The problem is how to safely thaw the twenty-three-ton carcass without destroying its tissue and organs. DNA can survive at temperatures of minus twenty-two degrees Fahrenheit or lower. The ice cellar where scientists will thaw the animal is too warm to keep the DNA intact. If the cells are too damaged for cloning, the scientists may attempt to use frozen sperm from the carcass to impregnate an elephant.

Another alternative to consider if the DNA is damaged is to synthesize the DNA of an extinct species in a laboratory. In December 2004, researchers at the University of California, Santa Cruz, reconstructed a million-letter-long DNA sequence from the "Boreoeutherian ancestor," an animal that lived seventy-five million years ago and is the ancestor of most placental mammals. Using the DNA sequence and biological components from living species, geneticists may be able to recreate the extinct animal's DNA and possibly even resurrect the species.

And what about cloning people? Professor Lu Guangxin of China's Xiangya Medical College says she and her team of researchers have successfully cloned human embryos for research purposes since 1999.

In December 2002, newspapers carried the story of a company, Clonaid, claiming that it had successfully produced a seven-pound baby girl cloned from the skin cell of a thirty-one-year-old American woman. But the story was never substantiated, and scientists doubt that the event occurred.

Colonies in Space

SCIENCE FICTION WRITERS envisioned humans colonizing outer space decades before the Russians sent *Sputnik* into orbit.

Probably the earliest fictional account of a human colony on another planet is *Edison's Conquest of Mars* written by Garrett Serviss in 1898. Serviss wrote the novel in response to H. G. Wells' *The War of the Worlds*, published in the same year and built around the Martians' desire to conquer Earth. The Serviss book was quickly followed by another novel of space colonization: *The Struggle for Empire*, written by Robert Cole in 1900. Since then, dozens if not hundreds of science fiction stories, novels, and films have portrayed human colonies on other planets or on giant spaceships traveling throughout the galaxy.

These include:

- J. B. S. Haldane's "The Last Judgment" (1927).
- Olaf Stapledon's *Last and First Men* (1930).
- Miles J. Breuer's and Jack Williamson's "Birth of a New Republic" (1930).
- Henry Kuttner's and C. L. Moore's *Fury* (1947).
- Eric Frank Russell's "...And Then There Were None" (1951).
- Isaac Asimov's "The Martian Way" (1952).
- James Blish's *The Seedling Stars* (1957).
- Harry Harrison's *Deathworld* (1960).
- Robert Heinlein's *The Moon is a Harsh Mistress* (1966).
- Ursula K. Le Guin's *The Left Hand of Darkness* (1969).
- Poul Anderson's *Tales of the Flying Mountains* (1970).
- Arthur C. Clarke's *The Songs of Distant Earth* (1986).
- Pamela Sargent's *Venus of Dreams* (1986).
- Paul J. McAuley's *Of the Fall* (1989).
- Kim Stanley Robinson's *Red Mars* (1992).

The idea of colonizing outer space generally takes one of two forms. The first is colonies that are literally "in space": humans living aboard vessels that are either traveling through space or orbiting a planet or star. The other notion of colonization is for humans to establish bases, and eventually towns and cities, on other planetary bodies, such as the Moon, Mars, or a planet in another solar system or galaxy.

For decades, the U.S. federal government has been talking periodically about space exploration and colonizing outer space. As science writer Timothy Ferris points out, "If we take the long view, the ultimate goal of manned space exploration is to establish permanent homes for humanity elsewhere in the solar system."

Space shuttles were designed to be quick, easy transport from Earth to outer space and back—a first step in making space exploration and eventually transport to space stations and space colonies more routine. But it has never come to pass, and a space shuttle launch has become as big a production as launching a traditional rocket—not what NASA originally intended.

Space stations could be considered the first attempt to "colonize" outer space. So in a sense, outer space colonies have already been established. But so far, they've been fairly modest in scope. The location is relatively close to home and limited to an Earth orbit. And the crew size—the population of the orbiting colony—is only half a dozen people or so.

The U.S. space station *Skylab* was put into orbit in 1973 and was home to a three-person crew for eighty-four days. The Soviet Union placed its space station *Mir* into orbit in 1986. *Mir* was used extensively by various crews from Russia, the U.S., and other countries for thirteen years. Both *Skylab* and *Mir* have since fallen out of orbit and into Earth's oceans.

In 1949 science fiction writer Willy Ley wrote a nonfiction book about space stations, *The Conquest of Space*; the book was a collection of his articles from *Collier's*, written with rocket scientist Wernher von Braun. James Gunn used the book as a reference for his novel *Station in Space* (1958).

Now the Bush administration, in the face of significant opposition from a public weary of recession and weak economies, is talking about sending astronauts to Mars. Could that be the first step in colonizing another planet in our own solar system? New photographic evidence has convinced some astronomers that Mars may have had water flowing abundantly over its surface at some point, giving rise to the hope that water or ice may yet be trapped under the surface. Finding water on Mars would make the establishment of a large, long-term colony more viable.

Colonizing a planet becomes much easier when the planet has its own water supply. It solves the obvious problem of a drinking water supply. The water molecules—which, as every student knows, consist of two atoms of hydrogen and one atom of oxygen—can also be used to generate oxygen for breathing and hydrogen to fuel rocket ships going back and forth between the colony and Earth.

The rotation of Mars gives it a twenty-four-hour day similar to an Earth day, which is favorable for plant growth. Scientists also believe Mars may have geothermal reservoirs below the surface that could be used to supply energy.

Many NASA scientists credit Ray Bradbury with instilling within them the curiosity and desire to explore and colonize outer space in general and Mars in particular. In his classic novel *The Martian Chronicles* (1950), Bradbury describes what the first Mars colonists would be like:

Mars was a distant shore. And the men spread upon it in waves. Each wave different, and each wave stronger.

The first wave carried with it men accustomed to spaces and coldness and being alone, the coyote and cattlemen, with no fat on them, with faces the years had worn the flesh off, with eyes like nail heads, and hands like the material of old gloves, ready to touch anything.

Mars could do nothing to them, for they were bred to plains and prairies as open as the Martian fields. They came and made things a little less empty, so that others would find courage to follow. They put panes in hollow windows and lights behind the panes.

Of course, one can argue that if we're going to colonize another planetary body in our solar system—a costly, ambitious, and risky undertaking—it might make sense to start closer to home with the Moon. After all, we've already proven our ability to send people there and get them back safely.

By one estimate, the Moon has ten billion metric tons of water split between its north and south poles. Another advantage is that the Moon is only a three-day trip by rocket from Earth, so emergency supplies and personnel could be transported quickly.

For a colony on another planet to be self-sustaining, it would need an energy source. If the planet cannot provide enough energy through solar collection or geothermal reservoirs, a small nuclear reactor could be built.

Colonists would have to erect structures for housing and community living, as well as a greenhouse for raising plants. The plants would provide food as well as generate breathable oxygen. You would have to wear a space suit any time you ventured outside the colony buildings. Raw materials for constructing the buildings—metals and minerals—could possibly be mined from the planet's surface, if they are present in sufficient quantities and accessible via current mining methods. Astronomers believe the large rocks of the asteroid belt orbiting the Sun between Mars and Jupiter may also be rich in metals and mineral resources. Robert Heinlein's 1952 novel *The Rolling Stones* is about a family that has a mining operation in the asteroid belt.

In addition to the obvious difficulties—food and water supply, energy, raw materials—colonization of outer space has at least one other problem: low gravity. The gravity on the Moon is one-sixth of Earth gravity, and of course there is zero gravity aboard a spaceship or space station. Without gravity, humans quickly lose muscle tone, and other long-term health problems might also occur. In a zero-gravity environment, we lose about half a percent of the calcium in our bones for every month in weightlessness. The danger is developing brittle bones that break easily, often a result of osteoporosis.

One solution is to create an artificial gravity. In *Star Trek*, the *Enterprise* seems to have an artificial gravity—the crew isn't floating around or wearing magnetic boots—but they really don't talk about it in the episodes. Space stations could be built to rotate at a speed such that the centrifugal force is enough to match or come close to Earth gravity. Colonists living on Mars or the Moon would probably have to contend with low gravity and all its consequences.

The farther from Earth the space colony, the more independent the local government could be: it would be difficult for an Earth government to enforce laws and regulations on colonists who are millions of miles away. The politics of colonization are raised in Edmond Hamilton's "Conquest of Two Worlds" (1932) and Robert A. Heinlein's "Logic of Empire" (1941); later in Robert Silverberg's *Invaders from Earth* (1958) and *Downward to the Earth* (1970), as well as Ursula K. Le Guin's *The Word for the World Is Forest* (1972).

Just how far from Earth could colonies be established? It depends on whether faster-than-light travel can ever be accomplished (see "Faster-than-Light Travel").

If we are restricted to the speeds of conventional solid or liquid fuel rockets, then aging of the astronauts puts a limit on the distance of the planets we can colonize. Even traveling at the speed of light, a twenty-year-old astronaut who leaves Earth to colonize a planet fifty light years away will be seventy when he gets there. One solution is to put the astronauts in suspended animation (see "Suspended Animation"), as was done in the original *Planet of the Apes* movie (1968), based on a novel by Pierre Boulle. In the film, Charlton Heston is horrified when, upon awakening from suspended animation, he sees that one of the suspended animation chambers has failed, and his fellow crew member has aged into a mummified corpse.

Another is to build "generational spaceships"—large ships with populations of families, including children who may live out their entire lives never having set foot on any planet. A generational spaceship arriving on a distant planet might colonize that planet with the grandchildren or even great-grandchildren of the original crew, all of whom have long since died.

Communications Satellites

O NE OF THE FIRST SCIENCE FICTION STORIES to deal with a satellite was probably Edward Everett Hale's *The Brick Moon* (1869). Joe Haldeman's "Tricentennial" (1976) takes place in part on a cylindrical satellite habitat.

Arthur C. Clarke, best known for his novels *Rendezvous with Rama* (1972) and *Childhood's End* (1953) among others, and for writing the motion picture *2001: A Space Odyssey* (1968), is commonly credited as the inventor of the communications satellite—an invention without which modern telecommunications would not exist. In a technical article (nonfiction) published in 1945 in *Wireless World* magazine, Clarke first proposed the concept of geosynchronous communications satellites. A geosynchronous satellite is a satellite placed in orbit at such a distance from the Earth that its orbital speed matches the speed of the Earth's rotation. Although it is in constant motion, the satellite is always positioned above the same spot on the Earth's surface.

In the article, he noted that it would take a minimum of three satellites to provide coverage to every location on the globe. He suggested positioning the first over Africa or Europe; the second over China or the oceans; and the third over the Americas at a longitude of ninety degrees west.

The title of Clarke's paper was "Can Rocket Stations Give Worldwide Radio Coverage?" He says, "It is the most important thing I ever wrote."

Writing articles for trade journals pays either nothing or a small honorarium; Clarke received the latter for his satellite-communications paper. He once commented that if he had only patented the design for a communications satellite network, he would have made billions of dollars.

Russia became the first nation to put a satellite into orbit, with the 1957 launch of *Sputnik*. In 1958, just thirteen years after the publication of Clarke's landmark paper, NASA began actively developing, building, and launching communications satellites.

Private industry entered the communications satellite business in 1962, when AT&T designed, built, and paid for the launches of its Telstar series of satellites. The original Telstar satellite weighed 175 pounds and was about the size of a beach ball. It was the first satellite capable of receiving and transmitting voice, television, and data signals across the span of the Atlantic Ocean.

NASA's satellite *Syncom 1*, launched in 1963, was the first satellite to achieve a geosynchronous orbit. But it failed before being put into use. Five months later, a second satellite, *Syncom 2*, also achieved synchronous orbit. It transmitted data, telephone, fax, and video signals. *Syncom 2* also had a wideband channel for television through which it provided coverage of the 1964 Olympics in Tokyo.

In a satellite communications system, each satellite carries transponders that operate at different frequencies. Each frequency corresponds to one or more receiving stations on Earth that are tuned to the same frequency. The communications satellite acts as a relay. Radio signals can be beamed up from Earth to the satellite, and then beamed down to another location.

Without communications satellites, two locations on the Earth's surface can communicate via radio signals only if there is a direct *line of sight* between them. Otherwise, the atmosphere, obstacles such as buildings, hills, and mountains, and even the curvature of the Earth block the radio waves from getting through: radio waves cannot reach over the horizon. Antennas for radio stations are located on tall towers to increase the coverage area of the broadcasts.

Sometimes, radio waves are bounced off the ionosphere to extend the reach of communications systems. But this method is subject to interference by atmospheric conditions. The solution is to place an object in the sky—a satellite—and bounce the radio waves off this man-made receiver. From a height above the Earth, a satellite can reach many more locations than a ground-based transmitter: the higher the satellite, the better its "bird's eye view" and reach.

At an altitude of approximately 186 miles, a height typically reached by a space shuttle, only 2.25 percent of the Earth's surface is reachable by the broadcast satellite at any one time. Go up to 23,500 miles, and a single satellite can cover 42.4 percent of the Earth's surface at once. The altitude of 23,500 miles is also the height at which a satellite keeps pace with the Earth as it orbits, always staying above the same spot on the planet's surface. Therefore, this is the altitude at which geosynchronous communications satellites are placed into orbit.

Today there are approximately 150 operating communications satellites in geosynchronous Earth orbit. Of these, thirty-six satellites, with
a total value of more than $4 billion, are provided by six private compa-
nies: GE Americom, Alascom, AT&T, COMSAT, GTE, and Hughes. The
satellites must come no closer to one another than one thousand miles,
or their signals will interfere with one another.

Communicators

I N SCIENCE FICTION MOVIES, the characters almost never call each other on the telephone: usually, they use some sort of modernistic communications device, a handheld communicator. *Star Trek* was probably the first movie or television show to actually discuss the communicators as part of the story: Crew members could not be located or beamed up when their communicators were lost or not working properly.

Dick Tracy, who made his debut in a 1931 newspaper comic strip, used his portable communicator on his wrist—where it doubled as a watch. And in the 1965 television show *Get Smart* Detective Maxwell Smart, an agent of Control, had his phone hidden in the heel of his shoe. Handheld communications devices have been available to the general public for decades. When I was a boy in the 1960s, many kids had walkie-talkies.

Now, with the advent of wireless communication, the handheld communicator is a reality, with millions of people worldwide owning cell phones, beepers, pagers, and personal digital assistants (PDAs).

As of this writing, the latest in portable communicators is the 3G, or "third generation," of wireless devices. Cell phones with built-in digital cameras allow you to take a color photo and transmit the image to the person you are talking with, as long as he or she has a picture phone, too.

With a 3G device on a wireless network, you can connect to the Internet without ever touching your computer. The 3G technology can transmit data at rates of up to three megabits per second. Wireless technology is also available for everything from laptop computers to video games: you can play Zelda or Tony Hawk without having to sit within a wire's length of your game system.

Cell phones aren't just "phones" anymore; they are morphing into full-fledged personal communicators that Captain Kirk would be proud to carry. T-Mobile's wallet-sized wireless Web device, the Sidekick, allows the user to transmit and receive e-mails, browse the Web, send documents or spreadsheets over the Internet as e-mail attachments, do instant messaging, take and transmit digital photos, and of course, talk on the telephone. It also has a miniature screen and keyboard which can be operated using your thumb.

Motorola sells a mobile handset, the i90c, which lets you communicate walkie-talkie style with compatible two-way radio phones. Another Motorola mobile handset, the 720, lets you send animated e-mail messages. Sony Ericsson's T300 mobile handset sends e-mail with text, animated graphics, and sound. Hewlett-Packard has announced plans to collaborate with Swatch to manufacture watches with wireless Internet connectivity.

The *New York Times* reports that Sprint is planning to introduce a mobile phone that can receive short downloadable video clips. Laptop and notebook computers have become increasingly lightweight; with wireless modems, they enable e-mail and even telephone conversation through VOIP (voice over Internet protocol) technology. The new handheld "palm top" computers are another type of personal communicator, with wireless access to the Web and e-mail.

Computers

FORWARD-LOOKING THINKERS imagined and experimented with computing devices centuries before the first electronic computers were built and tested.

Blaise Pascal, a seventeenth-century French mathematician, was the first to invent a computing machine. His "Pascaline" used a series of cogged wheels and gears to perform addition and subtraction; therefore it was really the first calculator, not the first computer.

In 1822, another mathematician, Charles Babbage, conceived of a machine that could perform much more difficult computations, including algorithms. He was the first to have the idea of "programming"—his computing machine would receive its instructions through punched cards ("programs"). It would also have a memory, capable of storing partial answers to be used in later computations. Babbage was able to convince the British government to provide the first round of funding for the construction of his computing machine. But the money ran out and the Babbage computer was never completed. The unfinished computer built by Babbage is on display in the Science Museum in London.

William Gibson and Bruce Sterling, two leading sf authors in the "cyberpunk" genre, ask "what if" Babbage has successfully completed his computing machine in their 1991 novel *The Difference Engine*. The book portrays England as controlling the entire planet by 1855 through cybersurveillance.

Pascal and Babbage were several steps ahead of the science fiction writers: the first story to feature a computer was Edward Page Mitchell's "The Ablest Man in the World" (1879). John Campbell wrote about benevolent computers in "The Metal Horde" (1930), "Twilight" (1934), and "The Machine" (1935).

In 1958, Isaac Asimov wrote a short story, "The Feeling of Power," about a handheld calculator capable of performing complex calculations. Texas Instruments was the first company to build and market sophisticated, lightweight, affordable handheld calculators capable of performing complex mathematical functions to high school and college students, scientists, and engineers in the 1970s. When these calculators first became available to students, parents worried that their children would become reliant on calculating machines and lose the ability to do math. In "The Feeling of Power," Isaac Asimov takes this idea to the extreme, imaging a society where people can't even add single-digit numbers without the aid of calculators.

In the 1950s, Arthur C. Clarke wrote *The City and the Stars*, in which an enclosed city is run entirely by a large computer, the Central Computer. It materializes whatever the residents need out of its memory banks.

In 1890, Herman Hollerith of MIT built an electromechanical device to process data stored on punch cards. He used his machine to tabulate results from the 1890 census in just six weeks. By comparison, the census of 1880 had taken nearly seven years to count by hand. The company Hollerith formed to build and sell his "tabulating machine" eventually became IBM.

IBM's early computers were massive by today's standards: in 1944, IBM constructed the first large-scale electromechanical computing machine, Mark I. A fifty-foot shaft helped coordinate the motion of 750,000 separate moving parts. On one particular occasion when the Mark I was not operating properly, the operators found that a moth had been trapped in one of the electromechanical relays, preventing it from closing. Ever since, problems with computer programming have been called "bugs."

The first true electronic computer, called the Electronic Numerical Integrator and Computer (ENIAC), was introduced in 1946. ENIAC took 1,800 square feet of floor space, was one hundred feet long, and weighed thirty tons. The computer contained eighteen thousand vacuum tubes, seventy thousand resistors, ten thousand capacitors, and fifteen hundred relays. Its main application: to compute firing and bombing tables for the U.S. Army. Back then, computers were programmed through hardware, not software: to get ENIAC to perform different functions, the computer operators literally had to rewire it. Computers have always been able to store the data they generated. In a 1945 paper mathematician Johnny Von Neumann suggested that the instructions for the computer could also be stored as data within the machine. His idea of a stored program greatly increased the computer's flexibility and ease of operation.

With the invention of the transistor at Bell Laboratories in 1947, computers began to shrink. Transistors—tiny chips of silicon with wires attached to them—began to replace bulky vacuum tubes. Computer processors were built at first on circuit boards and then on microchips, allowing them to dramatically increase in power and decrease in size. Today's laptops, which weight just a few pounds, have more computing power and speed than the early IBM mainframes.

In his 1959 story "The Martian Shop," Howard Fast made the first science fiction prediction of the miniaturization of computers, with a machine that fits into a six-inch cube. But he did not explain the miniaturization, nor did he predict integrated circuits.

Today computers continue to increase in speed and shrink in size. The driving force behind this is the continuous improvements in semiconductor manufacturing that permit a greater number of circuits to be fitted onto smaller and smaller microchips. Moore's Law, formulated by Intel cofounder Gordon Moore, says that the amount of processing power that can be contained on a computer chip doubles every 18 months.

Recently I attended a demonstration of a voice recognition word processing program at a trade show. Voice recognition technology has been around for several years now, although early versions of the software were prone to error. Users simply speak into a microphone, and the software translates the sound of their voices into a document that appears on the screen. Howard Fast described a calculator with speech recognition capabilities much earlier in his 1959 story, "The Martian Shop."

Science fiction writers have long worried that computers will become intelligent (see "Artificial Intelligence") and take control of the world away from humans, often when acting out of self-preservation. In E. M. Forster's 1909 story "The Machine Stops," all humans on Earth live in a worldwide city run by a massive computing machine.

The notion of computers turning against their creators like Frankenstein's monster began in 1926 with Edmond Hamilton's "The Metal Giants." Francis Bayer reversed this concept using a human revolt against computers in his 1941 novel *Tomorrow Sometimes Comes*.

In Fredric Brown's "Answer" (1954), a galactic computer network is completed, and when the switch is closed, is asked its first question: "Is there a God?" The computer answers: "There is now." Perhaps the most terrifying story of an all-powerful computer conquering humankind is Harlan Ellison's "I Have No Mouth and I Must Scream" (1967).

The complete literature of computers in science fiction is, of course, way too extensive to list. But here are just a few of the more famous and noteworthy early titles:

- Mark Clifton's and Frank Riley's *They'd Rather Be Right* (1957).
- Pierre Boulle's "The Man Who Hated Machines" (1957).
- James Gunn's *The Joy Makers* (1961).
- Gordon Dickson's "Computers Don't Argue" (1965).
- Frank Herbert's *Destination Void* (1966).
- D. F. Jones' *Colossus* (1966).
- Robert Heinlein's *The Moon is a Harsh Mistress* (1966).
- Martin Caidin's *The God Machine* (1968).
- David Gerrold's *When HARLIE Was One* (1972).

Cryogenic Preservation

CRYOGENICS, OR "CRYONICS," refers to the freezing of a human being—who is either near death or within minutes after death—at subzero temperatures to preserve the body. The idea is that the body can be revived and made healthy in a future when medical technology has advanced to the point where the subject can be cured of the disease that killed him. The term "cryonics" was coined by Karl Werner in the 1960s and popularized by Robert C. W. Ettinger in *The Prospect of Immortality* (1966). But the idea didn't catch fire with the public. When Frederik Pohl asked Ettinger how cryonics was going, Ettinger replied, "Many are cold but few are frozen."

The Cryonics Society of California began freezing people in 1967. Baseball great Ted Williams was one of them, causing some controversy among his children. There's a rumor that Walt Disney had his body frozen immediately after death and preserved in a cryogenic chamber, so that if medical technology develops to the point where it can cure the disease that killed him, he can be brought back to life.

As Thomas Slemen explains in his book *Strange But True* (1998):

> There are many sane and respected people around the world today who intend to have their bodies "put on ice" when they die. Their frozen corpses will be stored in liquid nitrogen at a temperature of -196 centigrade, until a future time when advances in medical technology will allow the deep-frozen dead to be resurrected.
>
> These attempts at cheating death through freezing are practical examples of the relatively young science of applied cryonics. The Cryonics Society of California is a pioneer in this field and started freezing newly-dead bodies in 1967. There are now cryonic storage societies starting up in other parts of the world.
>
> Many scientists still regard the prospect of cryogenic immortality as improbable and unlikely, as it is still difficult, if not impossible, to freeze human tissue fast enough to avoid vital-cell destruction. This

problem will undoubtedly be resolved in the not-too-distant future, and already, rudimentary human embryos have been successfully frozen at subzero temperatures. Moral watchdogs are concerned at the pace of progress in cryonics, and recent legislation in Britain has limited the period scientists can hold the embryos in cold storage.

An alien in John W. Campbell's 1938 story "Who Goes There?" (written under his pen name Don Stuart) comes back to life after being cryogenically preserved in ice at the South Pole for thousands of years, despite the fact that the scientists who find the body believe it to be impossible. One of them explains:

> You can't thaw higher life-forms and have them come to. A fish can come to after freezing because it's so low a form of life that the individual cells of its body can revive, and that alone is enough to reestablish life. Any higher forms thawed out that way are dead.
>
> There is a sort of potential life in any uninjured, quick-frozen animal. But it can't become active...though the individual cells revive, they die because there must be organization and cooperative effort to live. That cooperation cannot be re-established.

The first fictional use of cryonics was W. Clark Russell's *The Frozen Pirate* (1887) followed by Louis Boussenard's *10,000 Years in a Block of Ice* (1889) in which a contemporary man visits the future. Other sf works that use cryogenics as a means of visiting the future include Frederik Pohl's *The Age of the Pussyfoot* (1969), Mack Reynolds' *Looking Backward, From the Year 2000* (1973), and Woody Allen's film *Sleeper* (1973).

Marvel Comics superhero Captain America, who made his comic book debut in 1941, is cryogenically preserved when he falls into frosty arctic waters and is encased in a block of ice. Years later, his body is discovered and when defrosted, he comes to life. The supersoldier serum, a chemical compound that gave Captain America his strength, is credited with preventing his cells from crystallizing while he was suspended in a frozen state.

Sylvester Stallone is put into cryogenic suspended animation in the motion picture *Demolition Man* (1993). During his cold sleep, he is mentally reprogrammed to be less violent, without apparent success. He wakes up in a future so drab and uniform that Taco Bell is the only restaurant left in existence, the winner of the "franchise wars."

Other science fiction stories with a cryogenic theme include Clifford

Simak's *Why Call Them Back from Heaven* (1967), Ernest Tidyman's *Absolute Zero* (1971), and A. A. Attanasio's *Solis* (1994), in which cryogenically preserved corpses are revived to serve as cheap labor. Frederik Pohl and Larry Niven coined the term "corpsicle" to describe people whose cryogenically frozen bodies are mined for spare parts—Pohl in *The Age of the Pussyfoot* (1969) and Niven in *A World Out of Time* (1976).

But how scientifically feasible is it to revive the frozen dead?

In cryogenic preservation, liquid nitrogen or another cryogenic liquid (helium or argon can also work) is used to freeze and preserve the body at the instant of death. Several companies offer long-term storage of cryogenically preserved bodies, with their customers hoping that, like Walt Disney, they can be unfrozen and revived in the future once medicine has found a cure for the illness that killed them.

No one doubts the ability to preserve a human body at cryogenic temperatures; the technology has been around for years. But the obstacle to be overcome is reviving a cryogenically frozen living organism.

There appears to be some evidence that cryogenic suspended animation can work, albeit on lower life-forms. For instance, scientists found bacteria and algae in a salty Antarctic lake that has been frozen for more than 2,800 years. When the ice was thawed, the bacteria and algae came back to life. In Alaska, a species of bacteria was revived after lying dormant and frozen in the permafrost for thirty-two thousand years.

Human sperm, eggs, and embryos are routinely frozen for couples who may have future infertility problems because of illness or treatments. Thousands of these frozen embryos have been removed from cryogenic containers, thawed, and implanted in women who then gave birth to normal, healthy children.

Many different cells and tissues can be frozen and stored in liquid nitrogen and helium at temperatures hundreds of degrees below zero. In this frozen state, the cell's metabolic functions stop.

Certain species of fish and other animals have a natural "antifreeze" in their blood system that allows them to survive freezing temperatures. Known as antifreeze glycoproteins, these unique molecules prevent the fish from freezing in subzero waters by binding to and inhibiting the growth of ice crystals.

The eggs of the capelin, a fish related to smelts, can survive at temperatures as low as minus eleven degrees Centigrade. Barnacles can survive at temperatures of minus eighteen degrees Centigrade or lower, with more than eighty percent of their body water frozen.

A project, "Frozen Ark," is being launched to collect hair, skin, living tissue, and eggs from animals in danger of becoming extinct, and then freeze them at temperatures of 176 degrees Fahrenheit below zero. The hope is that, if an animal becomes extinct, these frozen samples can be used to bring back the species.

Bacteria and other microorganisms buried in ice can function at temperatures as low as minus sixty-eight degrees Fahrenheit. Microbes several million years old have been found in cryogenically induced suspended animation in permafrost at depths of almost two-and-a-half miles.

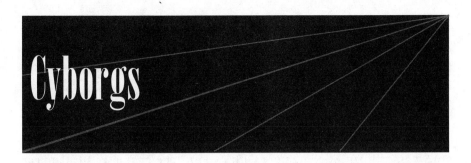

Cyborgs

THE TERM "CYBORG" was first used by Norbert Weiner in 1947. A cyborg is a being who is a mixture of organic and cybernetic (bionic) parts. The term *cyborg* is short for "cybernetic organism."

Kurt Vonnegut's *Player Piano* (1952) describes a cybernetic system. In "The Cybernetic Brain" (1950), Raymond F. Jones described human brains integrated into a computer, as do Frederik Pohl and C. M. Kornbluth in *Wolfbane* (1959), Chris Boyce in *Catchworld* (1975), and William Hjortsberg in *Grey Matters* (1971). Earlier examples of cyborgs in science fiction are E. V. Odle's *The Clockwork Man* (1923), Edmond Hamilton's "The Comet Doom" (1928), Curt Siodmak's *Donovan's Brain* (1942), and C. L. Moore's "No Woman Born" (1944).

The concept of turning a human being into a cyborg was proposed by Manfred Clynes and Nathan Kline in 1960 for space exploration. They envisioned a human altered to permit him to survive in extraterrestrial environments, as did Kim Stanley Robinson in his Mars series of novels. Cordwainer Smith's "Scanners Live in Vain" (1950) deals with men who are given mechanical parts to withstand the pain of space, as does Vonda McIntyre's *Superluminal* (1983).

One type of cyborg for space exploration is a human integrated with mechanical parts but still in a conventionally shaped humanoid body. The other cybernetic method of space exploration proposed by sf writers was to integrate a human (either the entire body or just the brain and central nervous system) directly into a spaceship. Examples include Thomas N. Scortia's "Sea Change" (1956), Anne McCaffrey's *The Ship Who Sang* (1969), Frederik Pohl's *Man Plus* (1976), and Gordon Dickson's *The Forever Man* (1986).

One can argue that any human enhanced with mechanical components is a cyborg. By that definition, a cardiac patient with a pacemaker is a cyborg. So is the person with an artificial leg. So is a pet dog with a chip implanted under his skin so he can be electronically tracked if missing.

Bernard Wolfe suggested that people might replace their limbs with prosthetics in *Limbo* (1952), as did Martin Caidin in *Cyborg* (1972), which was made into the television series *The Six Million Dollar Man* (see "Bionics").

To qualify as a true cyborg, you must be at least half machine; otherwise, you're just a person with some artificial parts. Examples of cyborgs include the Borg of *Star Trek: The Next Generation* and Officer Murphy, the robotic police officer in *Robocop*.

An assortment of characters with cyborg-type characteristics appear in William Gibson's 1984 novel *Neuromancer,* the first of the "cyberpunk" school of science fiction. One character, for instance, has her eyeglasses surgically inserted into her face, sealing her sockets. Retractable scalpel blades are housed in her fingers underneath artificial nails, reminiscent of the X-Man Wolverine and his adamantium claws.

In his novels *Man Plus* (1976) and *Mars Plus* (1994), Frederick Pohl suggests that augmenting humans by turning them into cyborgs could enable them to survive the hostile environments of other planets, such as Mars:

> Despite the triple redundancy built into his cyber systems and the constant checks they took with backup units orbiting overhead, Roger's computer controlled senses had become subject to intermittent failures.
>
> "Microseizures" he called them, when his world went black for two or three whole seconds while the backpack computer reset itself and then rebuilt his mechanical sensorium from the raw signals.
>
> Roger understood that he was just getting old. But what the actual design-life expectancy on his mechanical and cybernetic systems was, not even humans who built them could say. Alexander Bradley and the rest of the interface team back in Tonka had been shooting for a uniform fifty-year mean time between failures.
>
> That would have allowed Roger to live out at least the normal human span of three score and ten. As if he were normal anymore. Or, for that matter, human.

In Arthur C. Clarke's 1971 short story "A Meeting With Medusa," an astronaut who is injured when his rocket explodes during an expedition to Jupiter is rebuilt as a cyborg designed specifically for interplanetary travel and exploration:

> The life-support system inside the metal cylinder that had replaced his fragile body functioned equally well in space or under water. Grav-

ity fields ten times that of Earth were an inconvenience, but nothing more. Some day the real masters of space would be machines, not men—and he was neither.

And in the 1970s television series *The Six Million Dollar Man*, astronaut Steve Austin is turned into a cyborg when his damaged eye, arm, and legs are replaced with robotic substitutes (the show is based on the novel *Cyborg* by Martin Caidin).

In 2002, Kevin Warwick, a British professor, transformed himself into a partial cyborg when he had a three-millimeter silicon square surgically implanted through an incision in his left wrist. Surgeons attached one hundred electrodes from the chip into Warwick's median nerve. Connecting wires from the silicon chip protrude out of his skin. By connecting these wires to a transmitter/receiver device, Warwick has enabled his central nervous system to interface directly with a computer via radio signals.

Researchers at Georgia Tech have created the world's first cyborg rodent. The cyborg is controlled by two thousand neurons from a rat's brain. The robot is not connected to a rat or dissected rat brain. In the rat cyborg, a layer of rat neurons has been grown over an array of electrodes that pick up the neuron's signals. The neuron-covered electrodes serve as the "brain" of a three-wheeled silver robot. Sensors deliver feedback to the neurons through these electrodes. The robot moves in response to the neural activity of the rat brain cells. The neural signals from the rat brain cells are analyzed by a computer. The computer looks for patterns in the neural signals and translates these patterns into actions.

Deep Space Exploration

T HE IDEA OF SPACE TRAVEL in science fiction is hundreds of years old. Cyrano de Bergerac wrote about taking trips to the Moon and the Sun in a space rocket in his story "Voyages to the Moon and the Sun," published in 1662—more than four centuries ago.

Science fiction writers have been telling tales of space travel ever since. SF writer Jack Williamson calls space travel "the central myth of science fiction," much like the fall of Troy was the central myth of the ancient Greeks.

Edgar Allen Poe had men traveling to the Moon in a hot air balloon in "The Unparalleled Adventures of One Hans Pfaal" (1835), and Jules Verne sent them there by shooting them out of a cannon. H. G. Wells used a more practical spaceship, a sphere made of metal and glass, in his 1901 story "The First Men in the Moon."

After the Moon came exploration of the planets within our solar system, as in Garrett Serviss' *A Columbus of Space* (1909). Edgar Rice Burroughs used psychic powers to send John Carter to Mars in "Under the Moon of Mars" (1911), but used a spaceship in his 1930s Venus novels.

E. E. "Doc" Smith's *The Skylark of Space* (1928) was the first sf story to move into interstellar space exploration, along with Edmond Hamilton's "Clashing Suns" (1928) about an Interstellar Patrol. Others include:

- L. Ron Hubbard's *Return to Tomorrow* (1950).
- Robert A. Heinlein's *Have Spacesuit—Will Travel* (1958).
- Cordwainer Smith's "The Lady Who Sailed the Soul" (1960).
- Alexei Panshin's *Rite of Passage* (1968).
- Poul Anderson's *Tau Zero* (1967).
- Larry Niven and Jerry Pournelle's *The Mote in God's Eye* (1974).
- Frederik Pohl's *Gateway* (1977).

Two television series, *Star Trek* and *Lost in Space*, popularized the notion of space travel—not only throughout our solar system, but far beyond our own galaxy—to a broad audience beyond the traditional sf fan. Ray Bradbury regularly had astronauts visiting Mars in his stories, books, and TV scripts, most notably *The Martian Chronicles*. The notion of intergalactic travel was continued in such works as the Star Wars movie series and television's *Babylon 5*.

In 1969, U.S. astronaut Neil Armstrong became the first human being to set foot on the Moon. But reality has a long way to go before it catches up with science fiction, at least when it comes to space travel. We have sent unmanned spacecraft to Jupiter and beyond. But *manned* space travel has not gone beyond the quarter of a million miles from here to the Moon.

Launched by NASA in 1977, two *Voyager* space probes made it to Jupiter in 1979. Between them, they went on to explore, from a distance in space, Saturn, Uranus, and Neptune. They found a total of twenty-two new satellites orbiting these four planets. Some of these moons are thought to have surfaces of thick ice covering deep oceans of water. Certain scientists believe that these lunar oceans, not Mars, are the most likely places to find life on other planets within our solar system. Originally designed as a four-year mission to explore our solar system, the *Voyager* probes continue functioning today and are expected to have power and communication capabilities for another twenty years.

The *Voyager* space probes left our solar system in 1989, and are now headed toward the boundary zone, known as the heliopause, where the Sun's influence ends and interstellar space begins. A signal sent from Earth to the *Voyagers* traveling at light speed takes twelve hours to reach the spacecraft. But is deep space travel possible when people are aboard?

Consider our own solar system, which gives us a range of possible destinations, from Mercury closest to the Sun to Pluto farthest away. To travel these distances in a reasonable amount of time, you need a spaceship that can cruise at a speed of about one hundred miles a second or so. A nuclear-powered ship could possibly reach and maintain this velocity. So might a ship powered by a large solar sail.

What about visiting destinations beyond our own solar system? Even if we stay within our galaxy, the nearest star is about ten thousand times as far away as Pluto. To travel to such a destination within a human lifetime requires a spaceship that cruises at more than ten thousand miles a second—around five percent the speed of light—and can accelerate to this speed within ten years. Neither the government nor private industry currently has the technology to build a rocket ship that fast. Any engine today powerful enough to deliver that kind of thrust would quickly overheat.

To make interstellar space travel with a human crew possible would require one of the following:

1. A new rocket propulsion technology, such as the antimatter/matter "warp drive" on the U.S.S. *Enterprise* in *Star Trek*.
2. Discovery of a wormhole or other singularity in outer space that provides a shortcut through which a ship can circumvent ordinary space as we know it and "cut through" to reach its destination in a fraction of the time it would take in conventional space travel.
3. A cryogenic or other type of suspended animation system that would slow the aging process of the crew during the majority of the trip, during which time the ship would be on remote or automatic pilot.
4. A generational spaceship—a giant ship where families—even communities, of people—can live and procreate. Even if the velocity is modest by *Star Trek* standards, a crew consisting of descendants of the original astronauts will reach the destination eventually, barring mishap.

Robert Heinlein's "Universe" (1941) was the classic early statement of the generation spaceship. The concept was more firmly established in Murray Leinster's "Proxima Centauri" (1935) and A. E. van Vogt's "Far Centaurus" (1944). In George Zebrowski's *Macrolife* (1979), humankind travels through space in habitats made from hollowed-out asteroids.

Other Dimensions

PERHAPS THE EARLIEST novel-length treatment of alternate dimensions is William Hope Hodgson's *The House on the Borderland* (1908), in which the hero finds himself in a house that contains a portal to another dimension in which time moves at an accelerated rate:

> The world was held in a savage gloom. Outside, all was quiet. From the dark room behind me came the occasional, soft thud of falling matter—fragments of rotting stone. So time passed, and night grasped the world...wrapping it in impenetrable blackness.
>
> There was no night-sky as we know it. Even the few straggling stars had vanished, conclusively. I might have been in a shuttered room, without a light, for all that I could see. Only, in the impalpableness of gloom, opposite, burnt that vast, encircling hair of dull fire. Beyond this, there was no ray in all the vastitude of night that surrounded me; save that, far in the North, that soft, mist-like glow still shone.
>
> Silently, years moved on. What period of time passed, I shall never know...eternities came and went, stealthily; and still I watched. I could see only the glow of the moon's edge, at times; for now, it had commenced to come and go—lighting up a while, and again becoming extinguished.

Extension into other dimensions is featured in many sf works from the twentieth century:

- Miles J. Breuer's "The Captured Cross-Section" (1929).
- Donald Wandrei's "The Monster from Nowhere" (1935).
- Robert Heinlein's "And He Built a Crooked House" (1941).
- Martin Gardner's "No-Sided Professor" (1946).
- Arthur C. Clarke's "The Wall of Darkness" (1949).
- Bruce Elliott's "The Last Magician" (1952).
- David Duncan's *Occam's Razor* (1957).

In their 1941 short story "The Street That Wasn't There," Clifford D. Simak and Carl Jacobi explain that "in modern astrophysics and mathematics we gain an insight into the possibility and probability that there are other dimensions, other brackets of time and space impinging on the one we occupy." The premise of their story is that our world exists, unknown to us, in a dimension created collectively by the combined minds of all humans in existence. When a plague wipes out a large portion of the population, the dimension-creating power of humans as a species is diminished, until our reality gradually fades away. Since the story was published more than six decades ago, numerous additional science fiction stories have explored the idea of other dimensions. Now articles in respected science magazines are taking the possibility of alternate dimensions beyond the four we currently know.

One of the first writers to have a character fall into an alternate dimension was Lewis Carroll, whose heroine reached another dimension through a rabbit hole in his 1865 story *Alice's Adventures in Wonderland*. The Lost Boys in J. M. Barrie's *Peter Pan*, published in 1904, also exist in an alternate reality.

Mr. Mxyztplk, Superman's enemy from the DC Comics, is an imp or magical being who lives in another dimension. He frequently visits our dimension to annoy and pester the Man of Steel. The only way Superman can get rid of him is to trick him into saying his name backward, which temporarily returns him to his own dimension.

In the 1991 novel *Raft* by Stephen Baxter, the heroes are sent into another dimension where the gravity is much stronger. Now, some scientists are suggesting that gravity may indeed be leaking out of our own universe and into other dimensions (see "Antigravity").

In E. E. "Doc" Smith's 1949 novel *Skylark of Valeron*, the heroes briefly enter into the fourth dimension. A group of men in Clifford Simak's story "Hellhounds of the Cosmos" (1932) enter the fourth dimension to battle a monster.

In an episode of the original *Star Trek* television series, a transporter malfunction exchanges Kirk, Spock, and McCoy with their counterparts (evil, of course) in an alternate dimension. Philip José Farmer's protagonist Kickaha travels to alternate or "pocket" universes in his World of Tiers series of novels, the first of which is the 1965 novel *The Maker of Universes*.

In the 1970 Roger Zelazny novel *Nine Princes in Amber*, members of the royal family of Amber could move between dimensions through sheer willpower:

Something told me that whatever Shadows were, we moved among them even now. How? It was something Random was doing, and since he seemed at rest physically, his hands in plain sight, I decided it was something he did with his mind. Again, how?

Well, I'd heard him speak of "adding" and "subtracting," as though the universe in which he moved were a big equation.

I decided—with sudden certainty—that he was somehow adding and subtracting items to and from the world that was visible about us to bring us into closer and closer alignment with that strange place Amber, for which he was solving.

In the novel, Amber was the perfect world or dimension, and other dimensions—including the one in which Earth exists—were imperfect versions or "shadows." With their powers, the Amber royalty either moved to alternate dimensions or may have even created them.

Another explanation for multiple dimensions is that alternate realities are created every time someone makes a decision. For instance, if you decide to skip breakfast today, the decision creates two parallel dimensions: one in which you ate breakfast and another in which you skipped it. The universe in this way continually splits into parallel time tracks, a notion suggested by Philip José Farmer in his 1952 short story "Sail On! Sail On!":

> Parallel time tracks... it's his idea there may be other worlds in coincident but not contacting universes, that God, being infinite and of unlimited creative talent and ability, has possibly created a plurality of continua in which every probable event has happened.

Is the idea of another dimension pure fantasy, or might it have a basis in fact?

According to Einstein's General Theory of Relativity (1915), the space-time continuum has four dimensions. Three are the dimensions of space—length, width, and height—with time being the fourth dimension. Johann Zollner's *Transcendental Physics* (1881) used math to justify the "astral plane," and J. W. Dunne proposed a "Serial Universe" as a means of explaining prophetic dreams.

"String theory," a hypothesis that attempts to unify general relativity with quantum mechanics, says that the universe is made up of incredibly small strings vibrating in a space-time continuum consisting of eleven dimensions—the major dimensions of length, width, height, and time, plus seven "minor" (smaller) dimensions.

We live in a four-dimensional world, so we cannot perceive the other seven dimensions. The strings, however, vibrate in our four dimensions as well as the other seven, providing a connection between our dimension and others, and giving rise to the possibility of a parallel universe existing in those other dimensions.

Electric Cars

THE INFLUENCE OF AUTOMOBILES is dramatized in David H. Keller's "The Revolt of the Pedestrians" (1928), in which future drivers, having lost the use of their legs, depend upon broadcast power. Fritz Leiber's "X Marks the Pedwalk" (1963) features a war between drivers and pedestrians.

Today you can buy "hybrid cars" that have two sources of power. Typically, a conventional gasoline-powered internal combustion engine is the primary power source, with electricity-generating batteries the secondary power source.

But what about a truly electric car...one that has no gasoline engine and runs purely on electric current? In order for an electric car to be practical over long distances, the car needs to be equipped with a working fuel cell. Fuel cells, which run on hydrogen, have been used in the U.S. space program for more than three decades. For fuel cells to become a primary source of power for cars and trucks, their operation must be made much more economical. (Ironically, hydrogen, the most abundant element in the universe, is about five times as expensive per unit of usable energy as gasoline. The cost of power produced by hydrogen fuel cells is almost one hundred times as much per unit of power as that produced by a conventional auto engine. Yet ninety percent of all atoms in the universe are hydrogen atoms.) In a fuel cell, two metal plates—an anode and a cathode—are separated by a thin plastic membrane, or *proton exchange membrane* (PEM). The fuel is hydrogen, which consists of an electron orbiting a proton. When the hydrogen flow strikes the first metal plate, the anode, the electrons and protons are separated. The free electrons flow through a circuit as electric current. Meanwhile, the protons, stripped of their electrons, are combined with oxygen atoms in a two-to-one ratio, forming H_2O (water), which is exhausted from the fuel cell as its only waste product—thereby making fuel cells non-polluting.

The danger of fuel cells is that the fuel, hydrogen, is highly flammable: when exposed to a spark or flame, hydrogen explodes in a ball of fire. A gas stream will burn even with a hydrogen concentration as low as one part in fifty. Blimps originally used hydrogen to stay aloft until the explosion of the Hindenburg; after that, helium replaced hydrogen as the gas used in dirigibles.

Aside from the problem of safety, the other obstacle to making fuel-cell-powered cars practical is generating and handling the hydrogen fuel.

One common method is to use a hydrogen-containing compound as the fuel: the chemical reaction breaks the compound into its component molecules, and siphons off the hydrogen into the fuel cell. This can be done chemically on a small scale within the vehicle. For example, one fuel cell design uses sodium borohydride, a white crystal powder consisting of one atom of sodium, one atom of boron, and four atoms of hydrogen per molecule. The sodium borohydride is run through a catalyst which separates hydrogen atoms from the boron and sodium.

On a larger scale, hydrogen is routinely produced in large quantities during the refining of natural gas. Methane is the main component of natural gas, consisting of four hydrogen atoms bonded to one carbon atom. Large natural gas refineries are often equipped with a device called a steam reformer. The natural gas feedstock, which may also contain other light hydrocarbons, reacts with steam over a catalyst bed to produce "syngas"—a mixture of hydrogen and carbon monoxide. The syngas further reacts with steam in a converter, forming more hydrogen and converting carbon monoxide into carbon dioxide. The gas stream that emerges from the converter is approximately seventy-five percent hydrogen. Various technologies, including scrubbing and adsorption, are used to recover hydrogen from the syngas.

Electricity is often used to break water into its component parts of hydrogen and oxygen. The hydrogen can then be used in a fuel cell to generate power.

Researchers at the University of Minnesota recently produced hydrogen from ethanol in a prototype reactor small enough to install in an electric-powered car. The reactor could also be installed in a basement, where it could produce just about enough electricity to power an average home.

Storing hydrogen fuel is another problem: existing technologies require immense pressure and cryogenic temperatures to store hydrogen safely. Researchers at the University of Chicago have developed an alternate hydrogen storage method. They used a diamond anvil to compress crystals of hydrogen and water, and then cooled the crystals with liquid nitrogen. The result was a hydrogen-water "clathrate," or cage-like crystal that retained its 5.3 percent hydrogen by weight when returned to atmospheric pressure. When the crystals are warmed the hydrogen is released and can then be used by a fuel cell to generate power.

The Bush administration has budgeted $1.7 billion to develop the technology and infrastructure for generating and transporting hydrogen to be used in fuel-cell-powered vehicles and electrical power generation. One goal is to reduce U.S. dependence on OPEC oil through the increased usage of vehicles powered by fuel cells that consume hydrogen instead of internal combustion engines that run on gasoline or diesel fuel. Hydrogen fuel cells are about twice as efficient as internal combustion engines per unit of fuel consumed. Unlike internal combustion engines, which emit carbon monoxide and other pollutants, fuel cells give off only vapor and heat. The water produced by a fuel cell is clean and does not have to be treated before it is released.

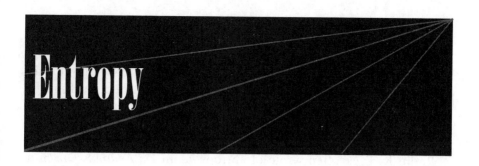

Entropy

IN A PURELY SCIENTIFIC SENSE, entropy deals with energy— more specifically, the lack or "unavailability" of energy in a closed system. Physicist Rudolf Clausius coined the term "entropy" in 1850 to describe the unavailability of heat or energy. Entropy is significant because it's the physical law that makes the ultimate demise or ending of the universe unavoidable. As Isaac Asimov explains it in his book *Understanding Physics* (1966):

> The second law of thermodynamics [states that] the total entropy of the universe is continually increasing. Now suppose the universe is finite in size. It can then contain only a finite amount of energy.
>
> If the entropy of the universe (which is the measure of its unavailable energy content) is continually increasing, then eventually the unavailable energy will reach a point where it is equal to the total energy. In the condition of maximum entropy, no available energy remains...The universe has "run down."
>
> Another manifestation of entropy is the degree of randomness or disorder in the universe: the tendency of systems (and the universe is the largest system of all) to degenerate from order into chaos. Since the second law of thermodynamics says that entropy is continually increasing, the universe is in essence "breaking down," transitioning from a state of order into chaos.

H. G. Wells describes a stage of entropy at the end of *The Time Machine* (1895): the time traveler has traveled far into the future, when our Sun has become a red giant.

In the DC Comics saga of Superman vs. Doomsday, Doomsday is so powerful a creature that even Superman cannot stop him. Doomsday is only destroyed when he is sent to the far future, where entropy has reached its maximum point and nothing (including Doomsday) can exist.

In a more poetic sense, entropy is used to denote a system or thing's decline or breakdown. Entropy became the dominant metaphor for the British "New Wave" science fiction (as opposed to American sf's metaphor of space travel) featured in Michael Moorcock's *New Worlds* beginning in 1964 and best characterized by Pamela Zoline's "The Heat Death of the Universe" (1967).

Philip K. Dick often wrote of entropy, as in *Do Androids Dream of Electric Sheep?* (1968), in which he calls it "kipple." Entropy is a major theme in such works as Thomas Pynchon's *Gravity's Rainbow* (1973), Robert Silverberg's "In Entropy's Jaws" (1971), and Michael Moorcock's *The Entropy Tango* (1981).

In Stephen King's Dark Tower series, the people in Roland the Gunslinger's plane of existence or dimension say that "the world has moved on"—expressing the feeling that things in the universe are slowly falling apart and breaking down.

Entropy as a metaphor for decline and decay is also prevalent in the short stories of J. G. Ballard—such as "The Voices of Time" (1960)—and is expressed in genetics: animals integrate lead into their shells to shield against radioactivity; the last fish alive on Earth struggles to survive in a shallow pond where a leaking dam threatens to spill the remaining water; and humans, unable to cope with the future, become genetically programmed, when genes within their cells become activated, to sleep almost twenty-four hours a day.

The most famous science fiction work about entropy (or at least with the word *entropy* in its title) is George Alec Effinger's novel *What Entropy Means to Me* (1972). In the book, the entropy in the universe is physically concentrated at a specific location, the "Well of Entropy." Here's how one character describes it:

> Down there is the entropic center of Home at least, and possibly the universe. Everything that falls down there becomes more and more dissociated, tending to the primal chaotic state. Matter is destroyed as the electrons constituting that matter slow and stop. When the electrons stop spinning and the atoms fall apart; no more matter, just randomly distributed, wasted energy.

Isaac Asimov writes about entropy in his 1956 short story "The Last Question," in which a supercomputer figures out a way to revive the universe after its heat-death by entropy. Here's Asimov's description of the universe approaching its end:

Man looked about at the dimming galaxies. The giant stars were gone long ago. Almost all stars were white dwarfs, fading to the end.

The energy once expended is gone and cannot be restored. Entropy must increase forever to the maximum.

When will the end finally come for our universe? Science writer Timothy Ferris says the universe will continue to support life for another one hundred billion years. If the universe continues to expand outward, it will eventually grow cold and dark as the stars run out of fuel and can no longer produce heat through fusion. As the universe expands, the density of matter within it becomes less and less, approaching but never quite reaching zero. Entropy increases, the amount of available energy (heat) decreases, and the temperature approaches but never quite reaches absolute zero.

This eventual scenario, billions of years into the future, is a cold, dark, empty expanse of space. Even matter will break down and fall apart as the degree of randomness within the system continues to increase toward its maximum. Molecules separate into atoms, and within the atoms neutrons, protons, and electrons separate into ions. Eventually, all that is left is empty space interrupted by an occasional ion floating around. This bleak scenario reminds me of a chilling line from the 1974 John Boorman motion picture *Zardoz*: "I have seen the future, and it doesn't work."

ESP (Psionics)

J. B. RHINE, a parapsychologist at Duke, popularized ESP in his 1934 book *Extra-Sensory Perception*. But does that mean people with psionic powers or extra-sensory perception really exist? Many people think so—and not just the lunatic fringe: a growing number of reputable scientists are taking seriously the possibility of mind powers.

Psionics ("psi" for short) refers to a whole range of mind powers including telepathy, the ability to read another person's thoughts; precognition, the ability to see into the future; telekinesis, the ability to move or manipulate matter with thought; teleportation, the ability to instantaneously move matter from one location to another; and pyrolisis, psychic fire-starting. ESP, short for extra-sensory perception, is a term used to describe a *set* of psi powers including clairvoyance, mind-reading, and precognition.

Early stories dealing with ESP include:

- Louis Tracy's *Karl Grier: The Strange Story of a Man with a Sixth Sense* (1906).
- J. D. Beresford's *The Hampdenshire Wonder* (1911).
- Stephen McKenna's *The Sixth Sense* (1915).
- Muriel Jaeger's *The Man with Six Senses* (1927).
- Edmond Hamilton's "The Man Who Saw the Future" (1930).

The first science fiction writer to use the term "psionic" was probably Murray Leinster in his 1955 story "The Psionic Mousetrap." Alfred Bester's *The Demolished Man* (1953) deals with trying to get away with murder in a telepathic society. In Roger Zelazny's *The Dream Master* (1966), a psychotherapist uses machine-enhanced psi powers to treat patients by manipulating their dreams during therapy sessions.

In his 1957 short story "Deeper than the Darkness," Harlan Ellison envisions a society where psionic powers are not rare, and people with them are grouped into categories. The "Mallaports" can manipulate matter—specifically, human flesh—at a local level, realigning the atoms in a material to change its composition or structure. "Drivers" have the ability to send themselves and everything around them into "inverspace" and are used to jump or "translate" starships through inverspace during interstellar voyages. "Blasters" can emit an energy field or beam with incredible destructive power. The main character of the story is an oddball—a "Pyrotic" who can start fires with his mind. The "Mindees" are telepaths.

Psi is an immensely popular topic with science fiction writers. Some notable examples include:

- Olaf Stapledon's *Last and First Men* (1930).
- James Blish's *Jack of Eagles* (1952).
- Henry Kuttner's *Mutant* (1953).
- Theodore Sturgeon's *More than Human* (1953).
- Wilson Tucker's *Wild Talent* (1954).
- Frank M. Robinson's *The Power* (1956).
- Lloyd Biggle, Jr.'s *The Angry Espers* (1961).
- John Brunner's *The Whole Man* (1964).

In Robert Silverberg's "(Now + n)(Now - n)" (1972), a stock trader gets rich from knowing which stocks will go up the next day by reading the mind of his future self. James Gunn's "The Reluctant Witch" (1953), Arthur Sellings' *Telepath* (1962), Lester del Rey's *Pstalemate* (1971), and Jack Dann's *The Man Who Melted* (1984) all deal with psi talents.

Stephen King's 1980 novel *Firestarter* is richly populated with characters who have psi abilities. Charlie McGee, the protagonist, can start fires with her mind. Her mother was mildly telekinetic, able to make doors open and close without touching them. Her father has a power he calls "push"—the ability to get others to do his bidding through mental influence.

In the 1960 movie *Village of the Damned*, based on a John Wyndham novel, aliens cause all the women in the village to become pregnant. Nine months later, they give birth to a group of children possessing psionic powers.

The most famous psi story may be A. E. van Vogt's *Slan* (1940), in which a new breed of humans with telepathic powers faces prejudice and discrimination from ordinary humans, a theme currently being explored in the X-Men series of movies. In Robert Silverberg's novel *Dying*

Inside (1972), a telepath loses his sense of identity as his telepathic powers slowly begin to lessen until he becomes an ordinary mortal.

Do ESP, telepathy, and other psi powers have any basis in scientific fact? The Nobel Prize-winning neurobiologist Sir John Eccles has suggested that a particle, which he calls a "psychon," carries thoughts. The brain stores thoughts as synaptic connections in the dendrites—the portion of the neuron that receives input. According to Eccles, when one thinks, these synaptic connections produce a psychon. If psychons could be transported or exchanged between individuals, telepathy, or mind-reading, would occur.

I recently interviewed famous psychic Uri Geller, who gained the spotlight by bending spoons with his mind on national television. Uri pointed me to a large section on his Web site (www.urigeller.com) filled with testimonials from scientists who either believe Geller has psychic power or at least insist that the power he has demonstrated cannot be explained by conventional science.

"I tested Uri Geller myself under laboratory-controlled conditions and saw with my own eyes the bending of a key which was not touched by Geller at any time," says Professor Helmut Hoffmann, of the Department of Electrical Engineering at Technical University of Vienna, Austria. "There was a group of people present during the experiment who all witnessed the key bending in eleven seconds to an angle of thirty degrees. Afterwards we tested the key in a scientific laboratory, using devices such as electron microscopes and X-rays, and found that there were no chemical, manual, or mechanical forces involved in the bending of the key."

In a test known as the ganzfeld experiment, the test subject must guess which of four images a viewer in another room is seeing. Probability says the test subject should guess correctly twenty-five percent of the time. In one set of ganzfeld experiments, 240 test subjects guessed correctly thirty-four percent of the time, a thirty-six percent increase over the expected amount.

U.S. intelligence forces report that between 1969 and 1971, the Soviet Union was engaged in psionic research. By 1970, it was suggested that the Soviets were spending approximately sixty million rubles per year on it, and over three hundred million by 1975. The money and personnel devoted to Soviet psi experiments suggested that they had achieved breakthroughs, even though the matter was considered speculative, controversial, and "fringy."

Researchers at Princeton University have run tests to find whether individuals can have telekinetic abilities, the ability to move matter with their minds. Specifically, the tests measure the level of *micro-psycho-kinesis*, the ability to influence microscopic events like the diffraction patterns of photon beams or the output of random number generators. Analysis of the results showed that some participants in fact exhibited this ability.

In the 1980s, the U.S. Government also conducted tests to determine the existence of *remote viewing*, the ability of a subject to know distant events through mind power. Many incidents indicating successful remote reviewing were documented by the researchers. Allegedly, a remote viewer "saw" that a KGB colonel caught spying in South Africa had been smuggling information in pocket calculator containing a communications device. The colonel was detained and questioned, and found to in fact have been a spy.

Exoskeleton

IONICS (see "Bionics" and "Cyborgs") enhance human strength and endurance from the inside out. An alternative way to boost physical abilities is to wear some type of hydraulically or otherwise powered device, either armor (an enclosed metal suit with the wearer inside) or an exoskeleton, in which the human wearer is only partially enclosed. Lobsters are born with an exoskeleton serving as a protective body armor, as are crabs and crayfish.

In Robert Heinlein's *Starship Troopers* (1959), soldiers wear mechanized suits that give them superior strength, vision, hearing, and protection against weaponry:

> Our suits gives us better eyes, better ears, stronger backs (to carry heavier weapons and more ammo), better legs, more intelligence (in the military meaning), more firepower, greater endurance, less vulnerability.
>
> The inside of the suit is a mass of pressure receptors, hundreds of them. You push with the heel of your hand; the suit feels it, amplifies it, pushes with you to take the pressure off the receptors that gave the order to push.

The Marvel Comics character Tony Stark invents an armor suit that gives him incredible powers of strength, invulnerability, flight, and weaponry, but also keeps his injured heart from beating. Wearing the suit, he becomes Iron Man, a member of the superhero team the Avengers. Iron Man made his comic book debut in 1963.

Bruce Sterling describes armor consisting of a wearable exoskeleton in *A Good Old-Fashioned Future* (1999):

> He'd come down from the heavens in his full NAFTA military power-armor, a leaping, brick-busting, lightning-spewing exoskeleton, all acronyms and elaborate gear.

In the film *Alien* (1979), Ripley wears an exoskeleton—a mechanical lifting device—to become a "temporary cyborg" in her fight with the acid-spitting alien on the loading dock of the spaceship. The human soldiers in the underground city of Zion in *Matrix: Revolutions* wear exoskeletons equipped with machine guns in the arms to battle the sentinels—machines that invade the city.

Now science fact is once again catching up with science fiction. Inventor Troy Hurtubise has built several versions of a personal suit of body armor designed to be strong enough to survive attack from a grizzly bear. His latest model, called the Mark VII, is made from stainless steel, heavy-gauge aluminum, and cast titanium. It features a built-in video screen, cooling system, pressure-bearing titanium struts, advanced protective airbags, shock absorbers, fingered hands, swivel shoulders, and built-in arms.

Yoshiyuki Sankai of the University of Tsukuba in Japan has developed a motor-driven, wearable exoskeleton called the Hybrid Assistive Limb, or HAL. By strapping HAL to the waist and legs, people with disabilities will be able to walk again. There is also an upper portion of the exoskeleton that enables the wearer to lift up to eighty pounds more than he or she could carry without wearing HAL.

Sensors attached to the wearer's skin detect signals sent from the brain to the muscles indicating an action to be taken, such as moving a leg. The signals are detected by the sensors, which activate the exoskeleton's motors and perform the action.

First Contact
(see also "UFOs")

$\mathbb{S}$OME OF THE IDEAS in this book are explored in science fiction infrequently; others are recurrent themes that have become staples of the field. Certainly "first contact" with other life-forms in general—and intelligent beings in particular—falls into that category.

Life on other planets was a common speculation, even among theologians, in the nineteenth century and later, and various humanlike creatures and Earth-like animals were encountered in Moon voyages and other extraplanetary journeys, dating back to Lucian of Samosata's "A True Story" in the second century A.D. and in Kepler's seventeenth-century "Somnium." Discussion of alien life may have been first popularized by Camille Flammarion's *Real and Imaginary Worlds* (1864) and his *Urania* (1889).

In J. H. Rosny Aîné's "The Shapes" (1887), aliens come to Earth in ships "a thousand years before the beginning of that center of civilization from which Babylon" was later to spring and terrorize a nomadic tribe, the Pjehu:

> It was first, a great circle of translucent bluish cones, point uppermost, each nearly half the bulk of a man. A few clear streaks, a few dark convolutions were scattered across their surfaces; each one had a dazzling star near its base.
>
> Farther, distant, equally strange slabs stood on end, looking rather like birch bark, and spotted with varicolored ellipses. Other Shapes; here and there, were almost cylindrical—some tall and thin, others low and squat, all of a bronzed color, topped with green; and all, like the slabs, having the characteristic point of light.
>
> Shapes began to sway in the twilight of the clearing. And suddenly, their stars wavering, flickering, the cones stretched higher, the cylinders and the slabs hissed like water thrown upon a flame, all of them moving toward the nomads with mounting speed.

Perhaps the most frightening alien depiction has been of the Martian invaders in H. G. Wells' *The War of the Worlds* (1898); the aliens were certainly less threatening in his *The First Men in the Moon* (1901). Wells was the first writer to bring the idea of aliens in general and Martians in particular to a broad audience.

Edgar Rice Burroughs' Mars and Venus novels featured humanlike men and women along with Earth-like animals and humanoid creatures with extra arms or legs and unusual colors. E. E. "Doc" Smith's aliens were generally bipedal and modeled after Earth creatures.

A significant departure from the "aliens will be a lot like us" model arrived with the aliens of Stanley G. Weinbaum, introduced in "A Martian Odyssey" (1934); these creatures were products of their environment and were hard to understand or interact with. Raymond Z. Gallun contributed to a depiction of aliens as not necessarily dangerous or deadly with "Old Faithful" (1934), followed by Ralph Milne Farley's "Liquid Life" (1936), Murray Leinster's "First Contact" (1945), Eric Frank Russell's "Dear Devil" (1950), and Arthur C. Clarke's devil-shaped Overlords in "Guardian Angel" (1950), which became the opening segment of his novel *Childhood's End* (1953). Other stories of contact with aliens include Fred Hoyle's *The Black Cloud* (1957), Orson Scott Card's *Ender's Game* (1977), and James Gunn's *Gift from the Stars* (2005), to name just a few.

Several novels deal with first contact by means of radio-telescope including James Gunn's *The Listeners* (1972) and Jack McDevitt's *The Hercules Text* (1986). The idea that radio telescopes should be listening for signals from extraterrestrial sources was proposed by Philip Morrison and Giuseppe Cocconi in 1959 and popularized by Walter Sullivan, science editor for the *New York Times,* in his book *We Are Not Alone* (1964). Paul Shuch says Gunn's *The Listeners* has done more to support the SETI (Search for Extraterrestrial Intelligence) program than any book ever written.

Olaf Stapledon's *Star Maker* (1937) showed aliens who were truly cosmic, not resembling humans in any way. John W. Campbell contributed a horrific shape-changing alien in "Who Goes There?" (1938, filmed twice as *The Thing*).

A. E. van Vogt's aliens in *The Voyage of the Space Beagle* (1950)—particularly the catlike alien first published in "Black Destroyer" (1939) and "Discord in Scarlet" (1939)—were said to have inspired the 1979 film *Alien*. Murray Leinster's "First Contact" (1945) suggested the need to cooperate with aliens, just as his earlier "Proxima Centauri" (1935) depicted the impossibility of compromise between flesh-eating plants and humans.

Another step forward in alien portrayals occurred in Hal Clement's *Needle* (1950), in which the aliens are viruses, and *Mission of Gravity* (1954), in which the aliens are caterpillar-like creatures adapted to life on a planet where the gravity is five hundred times that on Earth; Clement's aliens, like Weinbaum's, are believable products of their environment and conditioned by it. In Robert Silverberg's "Passengers" (1968), invisible, formless aliens terrorize humanity by temporarily taking possession of humans at will.

Stanislaw Lem offers a sentient alien ocean in *Solaris* (1961), recently made into a movie starring George Clooney. Many films reflect the dangerousness of aliens; one of the few that avoids this is *Close Encounters of the Third Kind* (1977).

Carl Sagan's book *Contact* (1985), made into a 1997 movie with the same name, is a "first contact" story. In the film, Jodie Foster constructs a device, based on plans transmitted to Earth by aliens, that's supposed to enable her to make first contact with the race of intelligent beings that sent instructions for the device's construction. She believes the device worked and transported her to a planet where she made mankind's first contact with an alien (played by David Morse of *Hack* and *St. Elsewhere*). But a videotape of the experiment shows she never left the device.

Arthur C. Clarke's *2001: A Space Odyssey* (1968) is perhaps the classic first contact story. In the movie, aliens accelerate the evolution of apes into Homo sapiens by using a mysterious black monolith to alter the thought patterns of a tribe of apelike creatures. The first act they learn: to use a discarded bone as a weapon and kill their enemy. Fast forward to the twenty-first century, where a manned expedition to the Moon uncovers, buried beneath the lunar surface, an identical black monolith (of course, they know nothing about the first). When uncovered and struck by the Sun's rays, this monolith sends a signal telling its alien creators that mankind has evolved to the point where he has achieved space travel—and that we are therefore ready for the next step in our evolution.

Most science fiction is optimistic about finding intelligent life in the universe, but Jack McDevitt's circa 1985 short story "Tidal Effects" is not: a scientist discovers that Earth is an "anomaly," and that all other Earth-like planets in the universe have hothouse atmospheres—similar to that of Venus—consisting of carbon dioxide; and all are incapable of supporting life as we know it.

In first contact stories, either we go to the aliens, or the aliens come to us. In the 1982 movie *E.T. the Extra-Terrestrial,* an alien botanist collecting plant samples on Earth is accidentally left behind by his crewmates. An Earth family finds and befriends him, naming him E.T. after a sound he makes when trying to speak. Eventually E.T. constructs a homemade communicator, notifies his ship of their error, and is picked up at the movie's end.

E.T. was cute, but the aliens who visit are not always friendly. In the M. Night Shyamalan motion picture *Signs* (2002), aliens come to Earth with the intention of using us as a food supply. Conveniently, they have an extendable appendage that sprays some kind of neurotoxin, rendering the victim unconscious and perhaps preserving the flesh for later consumption. Water turns out to be the kryptonite of this alien species; water burns them like acid, and they are forced to flee in the first good rainstorm since their arrival. Similarly, in the television series *Alien Nation,* the alien "visitors" who come to live with us on Earth are stronger than humans, but seawater burns them like acid.

A similar theme was central to the 1960s *Twilight Zone* episode written by Damon Knight, "To Serve Man." Giant aliens come to Earth offering to share with us the benefits of their advanced technology. They carry with them a book whose title, translated into English, is "To Serve Man"—reinforcing our belief that their mission is altruistic. Our perception alters when a linguistics expert translates the rest of the book. "To Serve Man—it's a cookbook," she tells a colleague, Lloyd Bochner, as he boards the alien ship for an exchange visit. Of course, he's there to be dinner.

During Orson Welles' famous radio broadcast of *The War of the Worlds,* many in the listening audience missed the disclaimer that they were about to hear a science fiction drama. As a result, thousands actually believed Martians were invading Earth. Turns out the Martians had no immune defense against Earth germs, and the attacking fleet is felled by the common cold.

So how well has science fact caught up with science fiction as far as first contact is concerned? As far as the legitimate scientific community—the one whose activities are tracked in *Science News* and the *New York Times* Science Section—is concerned, we have not yet proven that there is life in the universe other than our own, let alone intelligent life. UFOs (unidentified flying objects) are frequently cited, but the usual explanation given is that people are seeing a bright light, airplane, atmospheric disturbance, ball lightning, swamp gas, or other natural phenomenon. No UFO sighting has been confirmed to involve an actual spacecraft from another planet.

Several nonfiction books have been written claiming that an alien spacecraft crashed in Roswell, New Mexico, in 1947. The books also claim that the Government recovered the spaceship, the dead bodies of the alien crew, and one live alien (although it lived only a short time)— and that they have maintained a cover-up ever since.

The Roswell incident inspired the Men in Black (MIB) films, which posit that a small number of extraterrestrials have been living on Earth for decades. The MIB agents are responsible for policing the activities of these aliens and keeping the general public from learning of their existence.

There are scores of reports of aliens visiting Earth, but none verified by reliable news media, investigators, or scientific study. Thomas Slemen recalls one of the most famous reports of alien encounters on Earth in his book *Strange but True* (1998):

One of the best documented reports of a possible visitor from another world landing on Earth came from the little French town of Alencon, which is situated about thirty miles north of Le Mans.

At around five A.M. on June 12, 1790, peasants watched in awe as a huge metal sphere descended from the sky, moving with a strange undulating motion. The globe crash landed onto a hilltop, and the violent impact threw up soil and vegetation which showered the hillside.

A hatch of some sort slid open in the lower hemisphere of the globe, and a man in an outlandish, tight-fitting costume emerged through the hatchway and surveyed the observers with an apprehensive look. He started mumbling something in a strange language and gestured for the crowd to get away from him and his vehicle.

A few people stepped back, at which point the man ran through the break in the circle of spectators and fled into the local woods. Some of the peasants also ran away, sensing that something dangerous was about to happen and the remainder of the crowd decided to follow suit.

Seconds after the last members of the crowd had retreated from the sphere, it exploded with a peculiar muffled sound, creating a miniature mushroom-shaped cloud. The debris from the craft "sizzled" in the grass, and gradually turned to powder.

There have also been a number of reports from citizens saying that they were abducted by aliens. Alleged victims report being captured and taken aboard a spacecraft, where the aliens perform bizarre experiments upon them. The cartoon *South Park* spoofed the idea of alien experimentation, when aliens insert into Eric Cartman an anal probe that extends from his rear and opens up into a giant satellite dish.

In his book *UFO Sightings: The Evidence* (1998), Robert Sheaffer describes the most well-known of the alien abduction reports, the Barney and Betty Hill incident:

> Barney and Betty Hill were returning to their home in Portsmouth, New Hampshire, about midnight on September 19–20, 1961, after a vacation trip to Montreal. As they drove through a deserted area in the White Mountains of New Hampshire, they reportedly saw a star-like object that seemed to follow their car.
>
> When the object appeared to come closer, Barney Hill, getting out of the car to observe it through binoculars, believed that he saw alien faces peering at him through a row of windows. He ran back to the car, shouting to his wife that they were about to be captured; they drove off, frightened and confused and arrived home late.
>
> After they had been home for several days, Betty Hill began to have a series of dreams in which she envisioned that she and Barney had been taken aboard the supposed craft and given a physical examination by strange looking humanoid creatures.
>
> Several years later, when the Hills were undergoing psychiatric treatment, under hypnosis they each separately told of being "abducted" by alien beings during two supposedly "lost hours" in their journey—exactly as Betty had envisioned in her dreams.
>
> Official records reportedly have shown that numerous radar installations throughout New England tracked an unknown object landing and taking off at exactly the times stated in the Hills' account.
>
> Finally, under post-hypnotic suggestion, Betty Hill drew a so-called star map which she supposedly saw aboard the flying saucer. Reportedly these stars have all been identified by UFO researcher Marjorie Fish, and all turned out to be stars capable of supporting planets with life.

Is there life on other planets...and would it be similar to our own? More than one hundred planets have been detected orbiting other stars in solar systems within our galaxy. A team of astronomers recently located a solar system, about ninety light years from us, known as HD70642. It bears a striking resemblance to our solar system, with a planet about Jupiter's size, circling a star very much like our own Sun (approximately the same age and size), at about the same orbital distance as Earth from the Sun. The planet would therefore have a similar temperature to ours. But it has twice the mass of Jupiter, which would mean its inhabitants live on a world with a gravity far stronger than our own. If they came to Earth, they would be able to leap great distances and lift enormously heavy objects, much as Krypton's greater gravity helped give Superman super strength and the power of flight. In fact, most of the planets discovered are Jupiter-like in size and mass, making it likely that any aliens living on them would be far stronger than humans.

Of course, ninety million light years is a long way to travel, so any aliens visiting us from solar system HD70642 would not only be physically superior to us, but more technologically advanced as well.

In 1961, astronomer Frank Drake proposed that the probability of there being other intelligent life in the universe could be calculated by what is now widely known as Drake's Equation:

$$N = N_* f_p n_e f_l f_i f_c L$$

Where:

N = the number of civilizations in the galaxy capable of communicating with other civilizations

N_* = the number of stars in the galaxy

f_p = the percentage of these stars with planets

n_e = the number of planets with environments favorable to life

f_l = the fraction of planets on which life actually forms

f_i = the proportion of planets on which intelligent life evolved

f_c = the fraction of planets with civilizations able to communicate with others by radio transmission or other means

L = the longevity of the civilization.

In his book *Probability 1* (1998), Professor Amir D. Aczel shows, by applying statistics and astronomical data using Drake's Equation, that the probability of their being other intelligent life in our galaxy that we can communicate with is virtually one hundred percent.

With high-speed supercomputers processing an increased amount of signal data picked up by increasingly powerful radio telescopes, our ability to detect a signal from an extraterrestrial source is constantly increasing. Seth Shostak, senior astronomer for the Search for Extraterrestrial Intelligence Institute (SETI), predicts that we will make first contact with an intelligent alien species within twenty years. Using the

Drake equation, Shostak determined that there could be between ten thousand and one million civilizations in our galaxy advanced enough to send out radio broadcasts.

But do we—or the aliens—have to achieve interstellar space flight for first contact to take place? Yes, unless we can find life, intelligent or otherwise, right here in our own solar system. And it is Mars that has long been thought the likely candidate for finding life on another planet within our solar system.

On Earth, microscopic organisms called *methanogens* consume geothermal hydrogen as their fuel and give off methane as a byproduct. They are found only in hydrothermal water heated by vents on the floor of a hot spring, and they are the only living organisms on Earth that can survive in environments lacking organic carbon.

Animals respirate by taking in oxygen and expelling carbon dioxide, while plants do the opposite. Through photosynthesis, the plants collect solar energy and convert it into energetic electrons that provide power for biochemical tasks. Microbes living in thermal vents can oxidize and reduce geothermal compounds using hydrogen and carbon dioxide to form methane (the methogens)—or, they can use hydrogen sulfide and oxygen to produce sulfuric acid. No light is required. For instance, hot mud and a gas-laden plume of hot water are issuing from a volcano discovered on the sea floor between Greenland and Norway. Frozen methane hydrate—a white, ice-like solid with gas contained in a crystal lattice—caps the top. A new species of tube worm lives on the slopes, and bacteria may be trapped within the ice crystals.

Scientists theorize that methanogens might also exist on Mars and the Moon, which they believe to be devoid of organic carbon. Further, the evidence suggests that Mars once had—and may still have—a subsurface groundwater system and widespread volcanic activity to heat it and give off hydrogen for the methanogens to consume. Since the water is underground, no light reaches it, so only microbes that can survive without light could grow and flourish.

According to Vittorio Formisano of the Institute of Physics and Interplanetary Science in Rome, an instrument called the Planetary Fourier Spectrometer (PFS) has detected formaldehyde in the Martian atmosphere of 130 parts per billion (ppb). He believes the formaldehyde is being produced by the oxidation of 2.5 million tons of methane per year, more than could be produced by non-biological processes—and therefore, concludes that there must be bacteria-producing microbes living today in the soil of Mars.

In his book *The Monuments of Mars* (1987), Richard Hoagland makes the case that certain photographs of Mars' surface show what may be an ancient, abandoned city, which he calls Cydonia. If the shapes and structures of Cydonia turn out to be manufactured rather than natural, that would mean there was intelligent life on Mars at some point, whether native Martians or visiting aliens. Although skeptical that Cydonia was a city and not natural rock formations, the late Carl Sagan believed it was at least worthy of further serious investigation.

Other scientists believe microscopic living organisms may be found not in thermal vents on Mars, but in the oceans of the moons of Mars or Jupiter, with the Jovian moons Europa and Io being two possible candidates. Io is volcanically active. Several of the moons of Mars, Jupiter, Uranus, and Saturn are covered with ice under which astronomers believe there are vast, frigidly cold oceans. These include Jupiter's moon Europa, which has a highly fractured icy shell, and the smaller moons of Saturn: Mimas, Enceladus, Tethys, and Dione. The five major moons of Uranus are also composed mainly of ice and rocky material. On Earth, microbes have been known to thrive in temperatures as low as minus fifty-eight degrees Fahrenheit—well below freezing.

The space probe *Huygens* landed on Titan, Saturn's largest moon, in January 2005. Titan is larger than Mercury, has oceans of liquid ethane, and an atmosphere of nitrogen and methane. Some scientists believe that many of the chemical compounds that preceded life on Earth may be frozen on or below Titan's surface.

Photos sent back to Earth by *Huygens* of Titan's surface show deep drainage channels leading to a distant shoreline. Visible beyond it is either an empty basin or a dark lake. If Titan does have lakes or oceans, it would be the only world in our solar system other than Earth to have flowing liquid on its surface as well as a substantial atmosphere, two attributes capable of supporting life as we know it.

In addition, alien life may have already come to Earth—as a disease. Some scientists have proposed that the SARS epidemic may have started when pieces of a comet, contaminated with alien microbes, broke off the comet, fell to Earth, and landed in China where the SARS outbreak began.

One hundred tons of debris from outer space fall on Earth every day, most breaking up in the atmosphere. Some scientists believe that up to one percent of that material is bacteria. Particles carrying the SARS virus may have broken off the main body of the comet, been swept into the tail, and reached Earth as our planet passed through the comet's tail.

It may even be possible that life on Earth originated in such a fashion—microscopic organisms deposited on Earth by a comet's tail. So perhaps we have already met the aliens, and to paraphrase Pogo—they is us. The more likely scenario is that a comet colliding with Earth brought carbon, hydrogen, oxygen, and nitrogen to Earth, and these became the building blocks of life.

Flying Cars

WHEN I WAS A KID growing up in the 1960s, *The Jetsons* was a popular cartoon on TV. In it, George Jetson and everyone else in the world got around with flying cars.

I remember reading an article in *Popular Science* at the time showing how a real-life flying car could be built. It didn't look anything like George Jetson's car, which had a spaceship-like body, bubble dome, and a little jet in the back. The *Popular Science* version looked like a 1965 Chevy with airplane wings and a propeller bolted to the roof.

Many science fiction stories and films make use of flying cars, including the Star Wars series. Flying cars are featured in the films *Blade Runner* (1982) and *The Fifth Element* (1997), in which Bruce Willis drives a flying taxi cab.

The earliest sf story about a flying car is probably Jules Verne's 1904 novel *Master of the World*, in which the car not only converts into a flying machine but also into a boat and submarine:

> The deck and the upper works were all made of some metal which I did not recognize. In the center of the deck, a scuttle half raised covered the room where the engines were working regularly and almost silently. As I had seen before, neither masts nor rigging! Not even a flagstaff at the stern! Toward the bow there rose the top of a periscope by which the *Terror* could be guided when beneath the water.
>
> On the sides were folded back two outshoots resembling the gangways on certain Dutch boats. Of these I could not understand the use.
>
> In the bow there rose a third hatch-way which presumably covered the quarters occupied by the two men when the *Terror* was at rest.
>
> At the stern a similar hatch gave access probably to the cabin of the captain, who remained unseen. When these different hatches were shut down, they had a sort of rubber covering which closed them hermetically tight, so that the water could not reach the interior when the boat plunged beneath the ocean.

As to the motor, which imparted such prodigious speed to the machine, I could see nothing of it, nor of the propeller. However, the fast speeding boat left behind it only a long, smooth wake. The extreme fineness of the lines of the craft caused it to make scarcely any waves, and enabled it to ride lightly over the crest of the billows even in a rough sea.

As was already known, the power by which the machine was driven was neither steam nor gasoline, nor any of those similar liquids so well-known by their odor, which are usually employed for automobiles and submarines.

No doubt the power here used was electricity, generated on board, at some high power. Naturally I asked myself whence comes this electricity, from piles, or from accumulators? Unless, indeed, the electricity was drawn directly from the surrounding air or from the water, by processes hitherto unknown. And I asked myself with intense eagerness if in the present situation, I might be able to discover these secrets.

Suddenly a sharp noise was heard from the mechanism which throbbed within our craft. The gangways folded back on the sides of the machine, spread out like wings, and at the moment when the *Terror* reached the very edge of the falls, she arose into space, escaping from the thundering cataract in the center of a lunar rainbow.

There's a logical problem with flying cars that needs to be solved: air traffic control. Today a full-time team of air traffic controllers using sophisticated radar works at every major airport to keep planes from colliding. How would individual flying car drivers in a high-traffic area avoid collision? The velocity and proximity of flying cars in urban areas would be too great for individual drivers to prevent accidents, it would seem.

Inventor Paul Moller has been working on his "Skycar" for decades. In 1989, he built a model, the M200X, that has since made two hundred flights at heights up to fifty feet. It looks like a small flying saucer. Moller's latest Skycar model, the M400, can reach speeds of four hundred miles an hour and has a range of about nine hundred miles.

The Skycar M400 is powered by eight rotary engines in which a triangular rotor spins inside an oval-shaped chamber, creating compression and expansion as the rotor turns. If you're in an accident, airbags cushion you and a parachute slows your descent.

In 1990, Kenneth Wernicke built a small winged flying car that can be both driven and flown. Top ground speed is sixty-five miles per hour and air speed can reach four hundred miles per hour.

There are a number of obvious advantages to having a flying car. Aside from the sheer fun, you'd cut travel time considerably. Flying cars can reach speeds of four hundred miles per hour or higher, and they can travel the shortest distance between destinations—a straight line—rather than being restricted to routes determined by roads.

There are several disadvantages. One is the need for special driver's education to teach you how to drive a flying car, which is probably less complicated than piloting a plane but more complicated than driving a conventional auto.

Safety is also a factor: If two flying cars crash into each other at top speeds, you'd be having an accident going four hundred miles per hour rather than twenty to thirty miles per hour. Also, in addition to the impact, you'd fall twenty to sixty feet to the ground. To allow safe travel in flying cars, some sort of computer control would be required, such as an artificial intelligence-based navigation system (see"Artificial Intelligence") capable of reacting to changing traffic conditions in real time.

Fuel would not be exorbitant. Flying car prototypes run on gas, diesel, or even kerosene, and get decent mileage: about twenty miles to the gallon of gas. Once the flying cars go into mass production, the hope is that they could sell for as low as $60,000 base price.

In a less practical approach to building a commercially viable flying car, the mechanics on the TV show *Monster Garage* bolted 16-foot aluminum wings and a rudder to a 3,800-pound sports car, a Panoz Esperante. The car flew successfully with the wheels three feet above the ground for a distance of 300 feet—triple the distance of the Wright Brothers' first test flight.

Food Pills

CIENCE FICTION WRITERS and the media speculated a good deal about food pills throughout the early twentieth century. They were satirized in the film *Just Imagine* (1930). In his 1935 story "Alas, All Thinking," Harry Bates mentions food pills as a mechanism of the far future: humanity has evolved into a big brain on a pipe-stem neck, and atrophied bodies are fed by food pellets ejected into the mouth by a machine:

> The nasty little slit of a mouth under our host's head slowly separated until it revealed a dark and gummy opening; and as it reached its maximum I heard a click behind my back and jumped to one side just in time to see a small gray object shoot from a box fastened to the wall, and, after a wide arc through the air, make a perfect landing in the old gentleman's mouth!
>
> "He felt the need for some sustenance," Pearl explained. "Those pellets contain his food and water. Naturally he needs very little. They are ejected by a mechanism sensitive to the force of his mind waves."

In the 1973 movie *Soylent Green* starring Charlton Heston and Edward G. Robinson, real food was no longer available to anyone but the super-wealthy. The masses were fed on Soylent Green, a food substitute, which resembled a wafer more than a pill. The climax of the movie comes when Heston finds the truth about the wafers and warns the masses at the top of his lungs, "Soylent Green is made of people!"

The first "food pills" were developed in the 1960s for consumption in outer space by NASA astronauts. The food wasn't actually in pill form; it was a freeze-dried powder. You rehydrated it by adding water and drank it through a straw. A version of dehydrated orange juice, Tang, is sold in supermarkets today. Later on in the space program, the food substitutes came in tubes and were squeezed out like toothpaste. The oozing mess was so unappetizing, Gus Grissom snuck a corned beef sandwich aboard *Gemini 3*, for which he received an official reprimand.

Pillsbury developed a chewy "energy stick" that was first eaten by Scott Carpenter during his *Mercury* space flight (1962). Neil Armstrong and Buzz Aldrin also snacked on the Pillsbury energy stick during their *Apollo 11* mission to the Moon (1969). Today, you can buy many brands of energy bars at the supermarket or corner drug store.

The next step in food pills is micro-MREs, food tablets being developed that have enough calories in a single pill to sustain an active man or woman for 24 hours. The Department of Defense is also studying food patches that feed the wearer nutrients directly through the skin.

Canadian sf writer Robert J. Sawyer makes the logical case that food pills will never become a staple of the average person's diet because there is simply no need or advantage in eating them:

> Food is one of humanity's great pleasures, and in the future, it's only going to get better. Instead of changing foods, genetic engineering will change us, and the way we digest.
>
> Why should the foods we like best be the least good for us? Future generations of humanity will be able to eat any food, no matter how rich.
>
> Sugar, salt, fat, cholesterol—all the things we love but have to consume in moderation now will have no restrictions on them in future. All food will be nutritious; the sole criterion for choosing meals will be taste.
>
> And, of course, dishes that haven't been enjoyed for thousands of years will be back on the menu: we will resurrect mammoths and moas from recovered DNA.
>
> *Jetsons*-style food pills will never materialize; instead, in the future, enjoying sumptuous meals will be a guilt-free highlight of every day.

I have a sneaking suspicion that Sawyer is wrong and food pills may be around the corner. Already we can buy health bars in the supermarket or local drug store that substitute for an entire meal. Adults eat at McDonald's not because it tastes good or is good for you (it isn't), but because they're busy and need a meal they can eat while driving. A food pill fits this bill, and unlike a Big Mac, it can't drip on your lap.

Genetic Engineering

M AD SCIENTISTS IN SF FILM and literature love to tinker with the genes of innocent humans, animals, and even plants to either create perfect organisms with specific desirable qualities—such as strength, speed, intelligence, and the ability to resist extreme environments—or in some cases, just plain old mutated monstrosities.

In Aldous Huxley's 1932 novel *Brave New World* (see "Cloning"), genetic engineering was used to create standardized humans engineered for specific tasks and roles in society. Robert Heinlein used genetic engineering in *Beyond This Horizon* (1942) and Jack Williamson in *Dragon's Island* (1951). The 2002 *Star Wars: Attack of the Clones* featured a genetically engineered army manufactured for a private war.

Other sf works dealing with genetic engineering include:

- James Blish's *Titan's Daughter* (1961).
- Philip K. Dick's *The World Jones Made* (1956).
- Poul Anderson's "Call Me Joe" (1957).
- Damon Knight's *Masters of Evolution* (1959).
- Kobo Abé's *Inter Ice Age 4* (1959).
- Frank Herbert's *The Eyes of Heisenberg* (1966).
- Samuel R. Delany's "Aye, and Gomorrah" (1967).
- Ursula K. Le Guin's *The Left Hand of Darkness* (1969).
- John Varley's *The Ophiuchi Hotline* (1977).
- Charles Sheffield's *Sight of Proteus* (1978).
- Bruce Sterling's *Schismatrix* (1985).
- Brian Stableford's and David Langford's *The Third Millennium* (1985).
- C. J. Cherryh's *Cyteen* (1988).
- Robin Cook's *Mutation* (1989).
- Geoff Ryman's *The Child Garden* (1989).
- Nancy Kress' *Beggars in Spain* (1991).

But science fact has caught up with science fiction, and genetic engineering is an almost mundane reality today. As Nathan Aaseng explains in his book *Genetics: Unlocking the Secrets of Life* (1996):

> Thanks to the discovery of "restriction enzymes" in the 1970s, scientists discovered how they could use these enzymes to chip DNA into small pieces at precise locations. They then found other "recombinant" enzymes that attached DNA bits onto other DNA. With these tools, they could rearrange base pairs in strands of DNA, like editors splicing together a motion picture from several reels of film.
>
> Ananda Chakrabarty, a professor of biochemistry at the University of Illinois, took bacteria that lived on oil and used their genes to develop a new strain of bacteria that could break down oil into simpler substances.
>
> Pharmaceutical companies . . . have discovered how to insert a gene for insulin production into bacteria. Genes inserted in bacteria now pump out insulin cheaply and in great quantities as well as producing a variety of inexpensive chemical enzymes and other hormones.

In 1988 researchers at Harvard University created a genetically engineered mouse designed to be more prone to tumors than ordinary mice. Designed for cancer research, it was the first genetically engineered animal to be patented. Since then, more than three hundred new animal life-forms, created through genetic engineering, have been patented.

In 2000 scientists decoded more than ninety-seven percent of the genetic code of the fruit fly's 13,601 genes. That same year, Chinese scientists decoded more than ninety-nine percent of the genetic code of the *extremophiles* bacteria found in the Yunnan Province.

The $3 billion Human Genome Project, which started in 1989, has completed its initial goal of charting the entire genetic makeup of human cells. It has already helped geneticists identify many of the genes that cause cancer and other illnesses, as well as shown us which genes control various inherited characteristics. For instance, one gene identified recently, DBC2, produces a protein that suppresses tumor formation. The DBC2 gene was found to be missing or inactive in sixty percent of breast cancer patients. Researchers took samples from breast tumors in which the DBC2 gene was absent and grew them in test tubes. When the DBC2 gene was inserted into the tumors, they stopped growing and in some cases died.

Scientists today are using genetic engineering to create fantastic creatures. Lois H. Gresh and Robert Weinberg, in their book *The Science of Superheroes* (2002), describe one of these creations—a GFP (green fluorescent protein) bunny named "Alba":

> Alba is real, and was created by French genetic researchers through zygote microinjection. They removed fluorescent protein from a species of jellyfish, modified the gene so that it glowed more brightly, and then inserted the gene (called EGFG, the enhanced green fluorescent gene) into a fertilized rabbit egg cell that grew into Alba. The green gene was present in every cell of Alba's body.
>
> When Alba rested beneath a black light, the rabbit glowed green.
>
> The green gene did have important uses. It was used to code specific genes or proteins. When the protein was active, the fluorescence could be detected under a black light. In reality, scientists have used this tracing ability to observe anti-cancer genes by black light. In future, doctors hope to use the green gene method to help locate cancer cells in humans.

Today new developments in genetic engineering and gene therapy are being made at an accelerated pace:

- After being injected with muscle-building genes, a group of laboratory mice were able to run faster than—and almost twice as far as—ordinary mice. In addition, the genetic tampering prevented the mice from gaining weight, even when they were fed a diet high in fat.
- Researchers at the Hiroshima Prefectural Institute of Industrial Science and Technology have created genetically engineered silkworms that fill their cocoons with human collagen instead of silk. Collagen is used in weight loss products and other neutraceuticals. Other applications include drug delivery and tissue engineering.
- Genes that manufacture GABA, a vital neurochemical that calms motor movements of the body, are being injected into patients with Parkinson's, the disease afflicting actor Michael J. Fox and boxer Muhammad Ali. In an animal study with rats, researchers injected the gene into the brains of rats with chemically induced Parkinson's. Tremors were reduced in all of the test rats, and fifty percent showed significant improvement.
- Dr. Mark Brand of the University of Connecticut has inserted

frog genes into rhododendrons, creating in essence a hybrid plant/animal life-form. The objective is to give the bush an enhanced ability to resist root rot caused by the common soil fungus Phytophthora.

- Scientists have genetically altered roundworms to make them trim. The researchers identified four hundred genes in the worm's genetic code that control fat production and storage. By turning off three hundred of the genes, they have created slimmer roundworms—a technique that could lead to a gene-based weight loss treatment (about two hundred of the deactivated genes in the worms are also shared by humans).

- Dow has produced a strain of corn that can turn into biodegradable plastic. Other companies have inserted flounder genes into tomatoes to see if the fish gene can help the fruit stay fresh in cold weather. Pigs are being bred with human cells so that pig organs can be implanted in humans without being rejected by the human immune system.

- Mendel Biotechnology (named after Gregor Mendel, the discoverer of genetically inherited traits in living organisms) is developing a genetically modified rubber plant expected to yield five times as much rubber per acre of crop than natural rubber plants.

- Scientists at the Technical University in Munich, Germany, have identified the genes and three enzymes that control flavor in strawberries. They hope to genetically engineer strawberries to produce a sweeter, fruitier flavor, as well as a modified aroma. At the University of Munich, scientists have genetically engineered a strain of rice with more than twenty times the vitamin A of ordinary rice.

- University of Michigan researchers have developed a gene therapy that has restored hearing to deaf guinea pigs by stimulating the growth of new sound-detecting hair cells in their inner ears. This same gene therapy may some day be used to enable deaf humans to hear again.

Genetically Altered Food

ONE OF THE PRODUCTS of the genetic engineering revolution (see "Genetic Engineering") is genetically engineered foods.

Genetically altered food wasn't explicitly predicted by science fiction. In *The Food of the Gods* (1904), H. G. Wells wrote about a supernutrient, but it is not attributed to genetic alternation, a concept that just wasn't around at the time. And in Lewis Carroll's *Alice in Wonderland* (1865), Alice's body responds to a cake that says "eat me" and a bottle of liquid labeled "drink me," but the changes in size that are triggered are magic, not genetics.

But today, it's not magic that's creating supernutrients; it's generic tinkering in the laboratory. Since the Supreme Court ruled that life-forms could be patented for commercialization, thousands of genetically modified organisms (GMOs) have been patented. From 1997 to 1999, one quarter of American farm land—approximately seventy-five million acres—was converted to raise GMO crops. Today GMO ingredients are used in two-thirds of all processed foods in the U.S. Two million people eat products produced through biotechnology every day.

To create genetically engineered foods, ordinary plants are injected with bacteria, genes from other plants, and animal viruses. Such foods are called "transgenic" because the genes from one organism have been transferred into another. Often, transgenic foods do not taste different than conventional foods, and there is no way to tell you are eating genetically altered plant material. There are other reasons for genetic modification of corn, soybeans, squash, and other crops—in particular, resistance to weeds and pests.

To make transgenic plants, scientists first identify a gene that controls the trait or function they want to replicate. Using molecular tools, they reengineer the gene and incorporate it into the new organism (a plant).

Some crops are genetically engineered for nutritional value. For instance, Monsanto has produced a genetically altered mustard seed richer in beta-carotene content than natural mustard seed. Other plants are genetically engineered to transform them into "green factories"—plants that can produce antibodies, enzymes, and other therapeutic proteins.

Additional recent developments in genetically engineered food:

- Researchers at the government organization Fisheries and Oceans Canada have used genetic engineering to create salmon that grow more than seven times larger than ordinary salmon. And scientists at the Tokyo University of Marine Science and Technology have genetically engineered salmon that have the sperm-producing cells of a trout; the technique could eventually be used to breed endangered or valuable species, such as blue-fin tuna, considered a delicacy in Japan.
- Monsanto has created a genetically modified strain of corn that grows eleven percent more corn per acre, resulting in an added value of $24 per acre.
- Genes from the bacteria Bacillus thuringiensis has been engineered into new strains of corn and cotton, enabling them to produce crystal proteins that ward off insect pests. Thirty million acres of these crops are planted annually.
- Senesco Technologies, which specializes in delaying the aging and death of plants, has created a genetically modified banana with double the shelf life of natural bananas. The genetically modified banana ripens like a normal banana but does not develop black spots or spoil as quickly as regular bananas.
- Natick Labs, part of the Combat Feeding division of the U.S. Army, has numerous research projects underway designed to give soldiers portable food in condensed form that does not require refrigeration and has a long shelf life. Among the technologies they are experimenting with: carbohydrate-enhanced meat jerky, vacuum infusion of lipid-based suspensions into porous foods to boost the caloric content, and emulsification to generate a high-fat powder.

Although GMOs have a safe track record so far, with no incidents of disease or poisoning reported, some scientists worry about allergic reactions, toxicity, environmental risks, disease, and other health hazards. Some countries, including Japan and Australia, require food labels to state whether a food product contains genetically altered ingredients, but the U.S.—the largest producer of genetically altered corn and soy products—does not. Europe has banned import of genetically engineered food for several years.

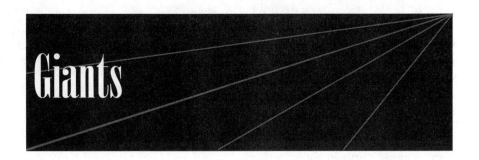

Giants

ONE OF THE MOST FAMOUS science fiction novels to feature giants was H. G. Wells' *The Food of the Gods* (1904), in which scientists experiment with making foods that promote growth at accelerated rates. Vermin get into the food supplies, grow to gigantic sizes, and terrorize the local citizenry. In the classic science fiction film *Them* (1954), the country is terrorized by giant ants.

Sometimes giants are humanoids. In Jonathan Swift's 1726 fantasy *Gulliver's Travels*, Gulliver travels to distant lands. In one, the people are miniature and he is the giant. In another, he is miniscule compared to the natives, all of whom are giants. In the 1968–1969 TV series *Land of the Giants*, the crew of a spaceship lands on a planet of humanoids of giant proportions. And in the Honey, I Shrunk the Kids film series, Rick Moranis invents a shrinking machine that, when operated in reverse, turns his youngest son into a giant. A woman grows to enormous proportions in the 1958 sf movie *Attack of the Fifty-Foot Woman*.

In real life, giants, albeit of smaller size than their science fictional kin, do exist, and gigantism is caused when the pituitary gland secretes an excess of human growth hormone.

Humankind has always been fascinated by very big people. When I was a kid growing up in the 1960s, my favorite basketball player was Wilt "The Stilt" Chamberlain, who was unique as the only NBA player whose height reached or exceeded seven feet. Today, seven footers in the NBA have become almost commonplace, and one player, Yao Ming, stands seven feet five inches tall.

The tallest human ever to have lived was Robert Wadlow, who reached a height estimated to be between 8'4" and 8'11 ½" as quoted by various sources. Wadlow's abnormal growth began at age two following a hernia operation. By age eighteen, he was 8'4" and wore size 37 shoes, specially made for him by International Shoe Company, who made the shoes free in exchange for Wadlow becoming their spokesperson.

Today the world's tallest man is 8'4" Leonid Stadnik, who lives in the rural Ukraine. Other modern giants include Hussain Bisad, who stands 7'9" and wears a size 26 shoe, and Chris Greener, who stands 7'6 ½" and is still growing. The world's tallest woman, Sandy Allen, is 7'7 ¼".

The size to which humans and other living beings can grow is limited, since the different areas of the body—bones, skins, etc.—grow at different rates. Giants often have difficulty walking because their skeletons cannot support their weight.

Can you create giants deliberately? Most definitely. In his book *How to Dissect* (1961), written for high school science students, William Berman describes a simple experiment for growing giant insects:

How would you like to try to produce a giant larva or a giant adult insect? It has been done! Several scientists have succeeded in producing giant moths. Let's see if the same experiment will produce giant grasshoppers. To try this you will need the following:

1. Several live young grasshoppers: you can keep them in jars filled with twigs and leaves, covered with perforated screw jar covers.
2. Five to ten abdomens of adult male moths. Moths may be used even if they are old and dried up, because the hormone in the abdomen remains unspoiled indefinitely.
3. A mortar and pestle for grinding the abdomens.

Begin your experiment by placing the moth abdomens in the mortar. Add ½ teaspoon of sand and grind the abdomens with the pestle until the mixture is like a powder.

Transfer the mixture to the beaker and add enough ether to reach a level of about 1 inch in the beaker. Now place in centrifuge tubes and centrifuge for 10 minutes. Pour liquid from centrifuge tubes on filter paper in funnel.

Wait until all the liquid comes through the filter paper into the vial. Place stoppered vial near open window to hasten evaporation of ether. The hormone will appear as a golden colored liquid.

Dip the paint brush into the hormone and paint a thin stripe on the abdomen of one of the young grasshoppers but do not cover the spiracles. Keep an untreated grasshopper in a separate container as a control, for comparison.

Keep accurate records of size and other visible changes. Now wait for results! Try this same experiment with other insects. Sooner or later the results will prove exciting.

While no one wants to deliberately create giants, doctors today use scientific methods to produce growth in children who are not growing at the same rate as their peers. By taking artificial thyroid hormone ("synthroid") pills or injections of human growth hormone (HGH), children who otherwise would have been tiny in stature have achieved normal size. HGH treatments do not create giants (though theoretically, the hormone could be used irresponsibly to do so). But it does allow people who would otherwise be small to achieve accelerated growth and height. In fact, children being treated with HGH often have to take calcium supplements; otherwise, their rapidly growing bones would not be strong enough to support their increased body weight.

Global Warming (the Greenhouse Effect)

I N HIS 1969 STORY "We All Die Naked," James Blish envisions a future Manhattan in which environmental changes render the air unbreathable (except with gas masks) and the entire city flooded because of rising tides from melting of the polar ice caps due to global warming.

In the 1970s film *Soylent Green,* global warming has transformed Earth into a greenhouse; it is always hot, and Charlton Heston sweats constantly. Air conditioning is the ultimate luxury, enjoyed only by the rich and powerful in this futuristic society.

George Turner wrote what is probably the most explicit global warming novel, *Drowning Towers* (1988), in which rising temperatures and ocean levels force nine-tenths of the population to live in high-rise towers. Frederik Pohl incorporated global warming into many of his stories and novels including *JEM* (1979) and *The Cool War* (1981). Global warming also takes place in J. G. Ballard's *The Burning World* (1964).

Unfortunately, scientists are increasingly convinced that global warning has made the transition from science fiction to science fact. They fear that man-made "greenhouse gases"—chlorofluorocarbons (CFCs), carbon dioxide, methane, water vapor, nitrous oxide, and other fumes that trap solar heat—in much the same way that a greenhouse warms the plants inside—are permanently raising the average temperatures around the globe.

In preindustrial times (circa 1750), the amount of carbon dioxide in the atmosphere was about 280 parts per million by volume (ppmv). Today it is about 370 ppmv—an increase of more than thirty-two percent since 1750, and up to ten times the levels of carbon dioxide present when dinosaurs walked the Earth during the Cretaceous period. In 1998, carbon dioxide emissions from U.S. chemical production totaled nearly 1.5 billion metric tons. To control atmospheric CO_2, researchers are developing a strain of plankton that can absorb carbon dioxide.

Here's how the greenhouse effect works: pollutants allow the Sun's rays to pass through the atmosphere. The Earth absorbs the solar radiation, then reflects the heat back. But in doing so, the wavelength of the radiation is changed. When the heat radiated from the Earth's surface hits the greenhouse gases trapped in the atmosphere, these gas particles absorb the energy and heat up. So in effect, they form a warm blanket of gas around the Earth, raising global temperatures.

Since we began keeping track in the late 1800s, the average world temperature has been 57.2 degrees Fahrenheit. In 1998, the global temperature rose to 58.51 degrees. Over the last 140 years, the average global temperature has risen by one degree Fahrenheit. Recent satellite studies predict overall global warming will continue at a rate of three-tenths of a degree Fahrenheit per decade. Scientists at the University of Oxford predict a more rapid global warming, estimating that Earth's temperature could rise by a whopping more than twenty degrees Fahrenheit if carbon dioxide levels double, as they are expected to do.

The more conservative estimate of just three-tenths of a degree may seem insignificant, but it is not. If greenhouse gas emissions continue unabated at their current level, then by 2070 the temperature increases could melt the polar ice caps sufficiently to raise sea levels and cause widespread flooding. Some scientists predict that by 2050 parts of England could be under water.

The Arctic is heating up twice as fast as any other region. Winter temperatures in Alaska and northern Canada have increased an average of five to seven degrees over the past half a century. A 386,100 square mile area of sea ice has melted in the past thirty years, and the remaining ice is ten percent to fifteen percent thinner. Global warming causes a portion of the Arctic ice cap the size of Texas to melt every ten years—which means we're losing nine percent of the ice cap per decade. At this rate, the entire Arctic ice cap will have melted by the end of the twenty-first century.

In the worst case scenario, a total meltdown of the ice in west Antarctica could raise the world's sea level more than fifteen feet, covering one-third of Florida with water. Currently the sea level rises two millimeters annually worldwide.

Global climate changes could also cause extreme weather events such as storms and hurricanes in some regions, while in other areas, rainfall declines and destroys crops. As Earth heats up, once-polar regions could become tropical, killing off polar bears, penguins, and other species unable to survive extreme heat.

The Kyoto Treaty of 1997 required industrialized nations to reduce worldwide emissions of greenhouse gases by an average of 5.2 percent below their 1990 level by 2007. But our own government seems indifferent to the crisis: in 2001, the U.S. pulled out of the Kyoto Treaty. And in October 2003, the U.S. Senate voted down a bill requiring reduction in emissions of greenhouse gases. However, six U.S. states and five Canadian provinces joined forces to combat global warming. They all agreed to reduce greenhouse gas emissions to 1990 levels by 2010.

In 2005, 141 nations ratified the Kyoto Treaty. But even if all countries meet their targets, they will only be able to eliminate one-tenth of a projected thirty percent increase in global greenhouse gas emissions between 1990 and 2010. Scientists from the National Center for Atmospheric Research say that if carbon dioxide emissions continue to rise, heatwaves in the U.S. and Europe will become more frequent, longer-lasting, and more intense, with minimum night-time temperatures almost six degrees Fahrenheit hotter than today's worst temperatures.

A new theory says that natural global warming may have already superheated the Earth once before. Approximately 250 million years ago, a period of intense volcanic eruptions and seismic activity might have released vast quantities of acid, sulfur, and carbon dioxide into the atmosphere, reducing oxygen levels from twenty-one percent to sixteen percent. The release of these greenhouse gases—and not, as previously thought, collision with an asteroid—raised the average global temperature by eighteen degrees, killing ninety percent of marine life and seventy percent of land animals.

Even our own solar system seems to conspire to make things hot for us on Earth: astronomers have discovered that, for the last quarter of a century, solar radiation reaching Earth from the Sun has increased an average of 0.5 percent per decade. And with the depletion of the ozone layer, more ultraviolet radiation gets through to us. We are literally being fried by the Sun.

Now NASA has come up with a new idea for combating global warming: create an artificial ring of small particles or spacecrafts around the Earth to shade the hottest areas of the planet from the sun's rays. The price tag: half a trillion dollars for the spacecraft or up to $200 trillion for the particles.

Holograms

NE OF THE EARLIEST science fiction novels to predict computer-generated illusions was James Gunn's *The Joy Makers* (1961), though it was not specifically described as a hologram.

In *Star Wars*, when Princess Leia is captured by Darth Vader and needs to send a distress call to Obi Wan for help, she records an audio-visual message featuring a holographic projection of her speaking—and has R2-D2 deliver it.

On *Star Trek: The Next Generation*, a favorite recreational activity aboard the Enterprise is a visit to the holodeck, in which the user can create a virtual reality of any time period using holograms.

What is a hologram? *The Concise Science Dictionary* (Oxford University Press) defines "holography" as a method of recording and displaying a three-dimensional image of an object using a laser and photographic plates:

> The light from a laser is divided so that some of it (the "reference beam") falls directly on a photographic plate. The other part illuminates the object, which reflects it back onto the photographic plate. The two beams form interference patterns on the plate, which when developed is called the hologram.
>
> To reproduce the image of the object, the hologram is illuminated by coherent light—ideally the original reference beam. The hologram produces two sets of diffracted waves; one set forms a virtual image coinciding with the original object position and the other forms a real image on the other side of the plate. Both are three-dimensional.

As in *Star Trek's* holodeck, holograms can be made to move. Specially filmed motion pictures, which must be viewed under a tungsten light, can be made using animated holographic images.

You may have seen 3D holograms displayed under a glass dome in museum exhibits. These three-dimensional objects look like ghostly images of the original, and can be viewed from all sides.

Holograms can depict anything from an inanimate object to a living animal or person. Normally movement of the subject distorts the hologram. But when making holograms of living organisms, a pulsed laser is used that emits a burst of light lasting just ten nanoseconds. The brevity of the light pulse allows holograms to be made of objects moving at speeds up to hundreds of miles an hour.

Although most holograms are decorative or artistic, some have commercial application. In partnership with a hologram company, Bo Dietl & Associates, a corporate security firm, offered a high-tech tracking service for apparel manufacturers who want to stop theft of their goods. Like an RFID (radio frequency identification) chip, the holograms—which contain data about the product's point of origin, distribution route, and intended market—are placed inside clothing labels. By tracking the holograms, the manufacturer can detect when shipments are being purloined.

Immortality and Longevity

NNATURAL LONGEVITY in literature has been around since the story of Methuselah, told in *Genesis* 4:21–27. The New Testament promises immortality to those who believe in Christ, but only after death. In her Anita Blake, Vampire Hunter series of novels, Laurell K. Hamilton writes of a church that guarantees members actual physical immortality: when you join, you are turned into a vampire.

Contemporary speculations often suggest that immortality can be achieved by downloading personalities into computers. A 2004 article in *Scientific American* seems to support such speculation, arguing that we are, in essence, the sum of the information encoded in our DNA and neural memories; therefore, the soul is nothing more than a pattern of information.

Any patterns of information can theoretically be duplicated on a computer—given enough storage capacity—implying that we can perhaps achieve immortality by duplicating this pattern in a storage device just prior to the death of our physical bodies. In Harlan Ellison's "Kiss of Fire" (1972), the protagonist frequently converses with his deceased wife, whose memories and personality are (albeit imperfectly) stored in such a computer memory device, which becomes flawed with age:

> It was an old memory box, the synthesizing channels were worn; the responses were frequently imprecise or non sequitur. The bead in which her voice had been cored had become microscopically encrusted; Annie now spoke with a slur and sometimes-drawl.

Classically, immortality has often been considered a curse, as in Sisyphus, Tantalus, Prometheus, and the Wandering Jew. Some horror novels suggest achieving immortality by becoming a vampire. Early apprehension-laden considerations of immortality include William Godwin's *St. Leon* (1799), Charles Maturin's *Melmoth the Wanderer* (1820),

W. Clark Russell's *The Death Ship* (1888), and Martin Swayne's *The Blue Germ* (1918).

Despite these pessimistic views of extended lifespans, humankind has been obsessed with staying younger looking, slowing the aging process, living longer, and of course, the notion of immortality—ever since Ponce de Leon (1460–1521) searched in vain for the legendary Fountain of Youth, and probably long before that. The Brothers Grimm describe a variation of the Fountain of Youth in their fairy tale, "The Water of Life":

> The Water of Life flows from a fountain in the courtyard of an enchanted castle; but you'll never get in unless I give you an iron rod and two loaves of bread. With the rod strike three times on the iron gate of the castle and it will spring open. Inside you will find two lions with wide-open jaws, but if you throw a loaf to each you will silence them. Then you must haste to fetch the Water of Life before it strikes twelve, or the gates of the castle will close and shut you in.

Longevity can be attained in several ways. The more mundane of them include exercise, nutrition, diet, and cosmetic surgery. The more exotic include mutation, chemical treatments, cloning of new bodies, genetic engineering, and other life-extension technologies, some of which promise immortality or close to it.

In Bram Stoker's *Dracula* (1897), people can achieve immortality by becoming vampires and drinking other people's blood. Now an experiment by researchers at Stanford University has established a link between blood and longevity. The researchers sewed pairs of mice together to surgically create conjoined twins with shared circulatory systems. When an older mouse was conjoined with a younger mouse, the older rodent became livelier and could regenerate muscle cells faster than older mice in a control group that were attached to other older mice.

Scientists speculate that a hormone in the blood of the younger mice causes cells to regenerate. Older mice have a diminished level of the hormone present, so their regeneration is slower. As the mice continue to age, the hormone disappears; the cells stop regenerating and the mouse dies.

In his 1958 short story "A Cross of Centuries," Henry Kuttner gives random mutation as the cause of immortality:

Immortality is an accident of the genes. A mutation. Once in a thousand years, perhaps, or ten thousand, a human is born immortal. His body renews itself; he does not age.

In Aldous Huxley's 1932 novel *Brave New World*, people are not immortal, but they are kept youthful well into their old age through advances in medicine:

We preserve them from diseases. We keep their internal secretions artificially balanced at a youthful equilibrium. We don't permit their magnesium-calcium ratio to fall below what it was at thirty. We give them transfusions of young blood. We keep their metabolism permanently stimulated.

Huxley also wrote about immortality in a less-famous novel, *After Many a Summer Dies the Swan* (1939). Immortality is a key element of Laurence Manning's *The Man Who Awoke* (1933) and George Bernard Shaw's *Back to Methuselah* (1921).

In his 1953 story "Whatever Happened to Corporal Cuckoo?," Gerald Kersh predicts that the formula for creating an immortality potion—which he called a "Digestive"—would be rather simple:

This fellow's brains were bursting out of his head. I applied my Digestive. And see what has happened. His eyes have opened! Observe, also that the bones are creeping together and over this beating brain a sort of skin is forming.

Ambroise Pare's Digestive, with which he treated the wounded after the Battle of Turin, was nothing but a mixture of oil of roses, egg yolks, and turpentine.

In his 1962 story *The Immortals*, James Gunn imagined a man, Marshall Cartwright, who, through some unexplained mutation, is born as the first immortal human. He passes on his immortality genetically to his heirs, but non-descendants can extend their lives almost indefinitely through periodic transfusions of the blood of either Cartrwright or his descendants.

In Clifford D. Simak's 1961 short story "Shotgun Cure," a visiting alien gives humans technology that makes them nearly immortal:

The alien said, "It does not really vaccinate. It makes the body strong. It makes the body right. Like tuning up a motor and making like new.

The motor will wear out in time, but it will function until it is worn out entirely."

Doc gingerly lifted out one of the pads and laid it on the desk. He kneaded it with a skittish finger and there was liquid in the pad. He could feel the liquid squish as he pressed the pad.

He turned it over carefully and the underside of it was rough and corrugated, as if it were a mouthful of tiny, vicious teeth.

"You apply the rough side to the body of the patient," said the alien. "It seizes on the patient. It becomes a part of him. The body absorbs the vaccine and the pad drops off."

"And that is all there's to it?"

"That is all," the alien said.

And in Arthur C. Clarke's 1963 story "Playback," aliens retrieve the essence of a human astronaut—his thoughts, intelligence, and memory—after his body is destroyed when his ship explodes, and contains it within a storage device, essentially granting him a type of immortality until a new body can be built for him:

I am a recording in some fantastic storage device. You must have caught my psyche, my soul, when the ship turned into plasma. All my memories are trapped in a tape or a crystal, as they once were trapped in the cells of my vaporized brain.

Another Clarke story published that same year, "The Secret," suggests that the reduced gravity of the Moon can greatly extend the human lifespan of lunar colonists, though not grant anything close to immortality:

On Earth we spend our whole lives fighting gravity. It wears down our muscles, pulls our stomachs out of shape. In seventy years, how many tons of blood does the heart lift through how may miles? And all that work, all that strain is reduced to a sixth here on the Moon, where a one-hundred-and-eighty-pound human weighs only thirty pounds. On the Moon, the span of human life will be at least two hundred years.

In Robert Silverberg's 1976 novel *Shadrach in the Furnace*, Shadrach is a surgeon whose main job is to replace failing organs in Khan, the ninety-three-year-old ruler of the world. Khan's researchers are conducting research projects designed to keep him immortal, one of which involves transferring his mind and personality into a younger body. Shadrach then makes the unsettling discovery that his body is the one Khan wants to inhabit.

Lazarus Long, the protagonist of Robert Heinlein's novel *Time Enough for Love* (1973), is over two thousand years old. Poul Anderson's 1989 novel *The Boat of a Million Years* follows the lives of a group of immortals over a period of thousands of years. Other sf works dealing with longevity or immortality include Jack Vance's *To Live Forever* (1956), Bob Shaw's *One Million Tomorrows* (1970), and Greg Egan's *Schild's Ladder* (2001).

While today we have not achieved immortality or anything close to it, we are certainly living longer than ever. In 1900, the average lifespan was just forty-nine years. Relatively few people made it into what we would consider today as "old age." As of 2002, the average lifespan for a man was seventy-seven.

Currently, just twelve percent of the U.S. population—approximately thirty-five million Americans—is age sixty-five or older. Over the next forty years, this will increase to twenty-one percent, and one out of five Americans will be sixty-five or older. By 2030, about seventy million Americans will be in the sixty-five plus age bracket. Researchers at the University of Illinois School of Public Health believe the average life span will reach eighty-five or ninety by 2100.

We are becoming an aged nation. The number of people older than one hundred in America has been increasing by more than seven percent a year since the 1950s. Today approximately fifty thousand Americans are one hundred years or older. By 2050, the number of centenarians in the U.S. is expected to reach 834,000.

Dr. Ronald Klatz, founder of the Academy of Anti-Aging Medicine, says that stem cell research will allow people to replace aging body parts. He believes people will eventually live to be at least 140.

Part of living long is simply good genes. Proof: one study shows that a man with a brother or sister who is one hundred is seventeen times more likely to reach that age than a man who does not have such a sibling. And if there are certain genes that contribute to longevity, those of us not born with them perhaps can gain them through genetic engineering.

One American, Henrietta Lacks, has achieved a kind of immortality. In 1951, a biopsy showed that she had cervical cancer, and she died eight months later. But the cells from her malignant tumor didn't. Normally, cells taken from human tissues kept in cell cultures outside the body can live for fifty generations (meaning they divide fifty times) and then die. But the "HeLa" cell line, grown from the living tumor of Henrietta Lacks, didn't die after fifty or even one hundred generations. They kept on growing, are still alive today over half a century since Lacks' death, and it seems as if they will continue forever. The sf version is Theodore Sturgeon's story "When You Care, When You Love" (1962), in which an individual is cloned from his cancer cells.

Professor Raymond Pearl, writing in the April 1921 issue of *The Scientific Monthly*, points out that while most of the body's cells divide continuously when in culture, and therefore are immortal, the human body made up of all these different cells is not:

> The potential immortality of all the essential cellular elements of the body either has been fully demonstrated, or else has been carried far enough to make the probability very great that properly conducted experiments would demonstrate the continuance of the life of these cells in culture to any definite extent.
>
> All the essential tissues of the metazoan body are *potentially* immortal. The reason that they are not immortal, and that multicellular animals do not live forever, is that in the differentiation and specialization of function of cells and tissues in the body as a whole, any individual part does not find the conditions necessary for its continued existence.
>
> In the body any part is dependent for the necessities of its existence, as for example nutritive material, upon other parts, or put in another way, upon the organization for the body as a whole. It is the differentiation and specialization of function of the mutually dependent aggregate of cells and tissues which constitutes the metazoan body which brings about death, and not any inherent or inevitable moral process in the individual cells themselves.

Some researchers believe the key to aging may be found in telomeres, small DNA-based structures at the end of chromosomes that shorten with each division of the cell. When the telomeres are gone, the cell stops dividing and of course, dies. If you can lengthen the life span of your cells, you will live longer. One theory speculates that by lengthening the telomeres, we should be able to increase the cell's lifespans by enabling it to divide many more times.

The key to lengthening the telomeres lies in the immortal part of Henrietta Lacks' body: cancer cells. In cancer cells, a protein called "telomerase" contains an enzyme component that helps build up and elongate the telomeres, so they lengthen with each division, instead of growing shorter as in noncancerous cells.

Through genetic engineering, a telomerase gene can be inserted into a normal cell. The cell will begin to produce telomerase while remaining noncancerous—and as a result, it will continue to divide, extending the cell's life span. Further research on the telomerase gene and the enzyme it produces may unlock the secrets of halting and reversing cellular aging in humans.

Another path to longevity may lie in a gene called daf-2. Researchers have found that by knocking out this gene, they have extended the lifespan of flies, mice, and worms. The gene controls numerous functions, leading some to speculate that it is a "master switch" through which the body programs aging.

More evidence that gene control can prevent aging was found by researchers at the National Cancer Institute, where skin cells taken from children with the premature aging syndrome Hutchinson-Gilford progeria (HGP) have been restored to normal. In HGP, a mutation causes a chunk of RNA to be cut out of skin cells when they replicate. Absence of this RNA segment produces an abnormal protein responsible for the early aging. The researchers "taped" the RNA segment in place using a short section of nucleic acid, preventing it from being cut out. Result: abnormalities in HGP patient skin cells were reversed.

Another key to longevity is P53, a protein that suppresses tumor growth. When a cell is damaged beyond repair or begins to behave abnormally, the P53 protein, whose production is controlled by the P53 gene, causes the cell to self-destruct, protecting against cancer.

Futurist Ray Kurzweil has identified three keys to increased longevity, and ultimately, immortality:

1. First, use all available orthodox and alternative medical knowledge to keep healthy (Kurzweil takes 250 dietary supplements daily, from alpha lipoic acid to milk thistle).
2. Benefit from new advanced in medical technology, such as genetic tests to determine predisposition to cancer.
3. Be an early adaptor of cutting edge technology. For instance, Kurzweil imagines that nanorobots could replace his digestive system, doing a more efficient job of extracting nutrients from food.

In her 1950 short story "And Be Merry," Katherine MacLean describes an experiment in which a scientist attempts to reverse the aging process by regenerating every cell in her body. For a time, the experiment seems to succeed; and she grows younger and younger as the story progresses. But some of the mutated cells grow out of control, and the cell regeneration treatment eventually gives her cancer.

And in his circa 1985 short story "Evergreen," Arthur Cox describes a technology that can prevent—at least for a time—cellular aging in human beings:

> From Nature's viewpoint there's no reason to hang around once your kids are old enough to take care of themselves, so a self-destruct mechanism is triggered which prevents the cells of our bodies from repairing themselves and so we get old and sick and die.
>
> Dr. Ivers discovered a way to defuse the self-destruct mechanism and this made it possible for people to remain in the price of life and no longer die of "old age," as it was called.

In Cox's story, the maximum lifespan attainable using this technology was approximately two hundred years. The reason: eventually, the natural genetic errors that sometimes occur in cell duplication (e.g., a cell replicates imperfectly, creating a nonfunctioning "offspring") accumulate over time to where enough of the body's cells cease to function normally, and the person given the Dr. Ivers' life extension process dies anyway.

There are too many additional immortality-themed sf works to list them all here. A few more key novels and stories worth mentioning: Damon Knight's "World Without Children" (1951), John Wyndham's *The Trouble with Lichens* (1960), Norman Spinrad's *Bug Jack Barron* (1969), and Mack Reynolds' and Dean Ing's *Eternity* (1984), to name just a few.

Some recent bad news in the quest for immortality: microbiologists have discovered that bacteria, once thought to regenerate indefinitely, eventually die, just like every other living creature. Subsequent generations in a colony of bacteria are smaller and more likely to die than their "parents."

Rabbi Harold Kushner, writing in his book *Living a Life That Matters* (2001), suggests that failing to achieve immortality might not be the tragedy for the human race that it seems.

"People don't really want to live forever," suggests Kushner. "Living forever would be like reading a good book, or watching a good movie, that never ended. People understand that the story of their lives has to have a beginning, a middle, and an end."

The Internet

$\mathbb{S}$ CIENCE FICTION WRITER John Brunner (1934–1995) is credited with having predicted the Internet. In his novel *The Shockwave Rider* (1975), characters surf continent-wide data networks and set loose "data worms" that hack information and systems.

Bruce Sterling and William Gibson, sf authors from the "cyberpunk" school, were among the writers who envisioned a "wired world" in which users could plug their minds directly into a data network and consciously experience and manipulate this environment.

For instance, Case, the protagonist of William Gibson's *Neuromancer* (1984), connects directly with Gibson's version of the Internet through electrodes implanted in his body:

> The moderns were using some kind of chicken-wire dish in New Jersey to bounce the link man's scrambled signal off a Sons of Christ the King satellite in geosynchronous orbit above Manhattan. Molly's signals were beamed up from a one-meter umbrella dish epoxy-ed to the roof of a black glass bank tower nearly as tall as the Sense/Net building.
>
> Case gulped the last of his coffee, settled the trodes in place, and scratched his chest beneath his black t-shirt. He had only a vague idea of what the Panther Moderns planned as a diversion for the Sense/Net security people.
>
> He jacked in and triggered his program. "Mainline," breathed the link man, his voice the only sound as Case plunged through the glowing strata of Sense/Net ice. Good. Check Molly. He hit the simstim and flipped into her sensorium. The scrambler blurred the visual input slightly.

Perhaps one day humans and data networks will interface directly. But today we can only access this interconnected data network through a PC or computer terminals.

No science fiction writer envisioned the current incarnation of the Internet, a worldwide network (the World Wide Web) in which every computer could immediately communicate with every other computer. J. C. R. Licklider of MIT came closest to predicting the Internet; in August 1962 he wrote a paper on what he called the "galactic network," a globally interconnected set of computers through which everyone could quickly access data and programs from any site. Licklider's galactic network more closely describes today's Internet than any sf novel or story.

The Internet began in 1969 as a computer network named "ARPANET" (Advanced Research Projects Agency Network). It was sponsored by the Pentagon so that scientists and researchers could share computer facilities and data and collaborate on projects.

ARPANET users—at first, mainly scientists and government employees—quickly began to use ARPANET as a messaging system to communicate news, memos, even personal news and gossip.

In the 1970s, the U.S. Defense Advanced Research Projects Agency (DARPA) began to develop the communications protocols that would allow networked computers to communicate transparently across multiple networks. A "protocol" is a set of standards and procedures that control the format and timing of data transmitted between two devices. For two devices to communicate, they must use the same protocol.

The original protocol for achieving what DARPA called "Internetting" was the Network Control Protocol (NCP). Within a few years, ARPANET upgraded to a newer, more flexible communications protocol, TCP/IP, which stands for Transmission Control Protocol/Internet Protocol. This is the protocol used by the Internet today. Hardware manufacturers began building routers, switches, hubs, and other networking equipment using the TCP/IP standards. By doing so, they allowed every TCP/IP-compatible computer to be connected to one another over ARPANET, which became known as the Internet.

How is data actually carried from your computer to another computer over the Internet? The data is routed using the TCP/IP protocol. TCP/IP ensures that streams of bytes (data) can be reliably communicated between two processors attached to an interconnected network.

But where is this interconnected network—the wires and cables that physically carry the data packets from one computer to another over the Internet?

In 1986, the U.S. National Science Foundation (NSF) developed NSFNET, the major portion of the communications network infrastructure over which Internet traffic is transmitted. NSFNET carries twelve billion packets of data monthly at rates of up to forty-five megabits per second between the computer networks in thirty-three countries. NASA also contributes a portion of the Internet's network infrastructure, called NSINET. The U.S. Department of Energy similarly contributes an additional portion of the Internet's network infrastructure called ESNET. Other portions are provided by commercial network providers in the U.S. and Europe.

Regional support for local nodes in the Internet, which receive traffic from the main network infrastructure and route it locally, is provided by a variety of federal and state government agencies, corporations, and universities.

While the various entities that run parts and pieces of the Internet are loosely affiliated and governed by an organization known as the Corporation for Research and Educational Networking (CERN), these entities are voluntary members. Therefore, there really is no central, single organization with responsibility for keeping the Internet up and running.

Can you imagine what would happen if, tomorrow, you turned on your computer and found the Internet was down or, worse, closed for good? Perhaps there is a science fiction story waiting to be written in which the individual entities running portions of the Internet go out of business or cease operations, and the Internet collapses.

There is a deliberate reason, however, why the Internet has evolved into a disparate collection of transmission facilities and local operators. In the 1960s, the RAND Corporation, famous as a think tank, was given the task of coming up with a system through which the U.S. authorities could continue to communicate even after a nuclear war. Their solution was to create a network with *no* central authority. The reason: If the network was centralized, and the location was ever discovered, an enemy would make the central network location a prime target in a military strike.

Invisibility

THE FIRST SCIENCE FICTION NOVEL to bring the notion of invisibility to widespread attention was H. G. Wells' *The Invisible Man* (1897). The plot is simple: a scientist invents a formula that, after he drinks it, renders him invisible. Because the chemicals alter only his body chemistry, the formula does not turn his clothes invisible, so he is forced to strip naked if he wants to be unseen. In the book he explains the principles of invisibility:

> I found a general principle of pigments and refraction—a formula, a geometrical expression involving four dimensions. It was an idea by which it would be possible, without changing any other property of matter—except, in some instances, colors—to lower the refractive index of a substance, solid or liquid, to that of air—so far as all practical purposes are concerned.

A modern version of this story is the 2000 film *The Hollow Man*, starring Kevin Bacon. In both the Wells novel and the Bacon film, invisibility has the unwanted side effect of making the scientist/hero mentally unstable, turning him homicidal.

In the James Bond film *Die Another Day* (2002), the latest version of the Bond car has a button that, when pressed, renders the car and its contents virtually invisible through cameras and video screens placed all over the car. In DC Comics, the Legion of Superheroes features a member, the Invisible Kid, who has the power to turn invisible at will—clothes and all—as does Sue Richards, the Invisible Girl from the Marvel Comics superhero team The Fantastic Four.

Now it appears that science fact has once again caught up with science fiction: researchers at the University of Pennsylvania claim that they have developed a special coating that can make solid objects nearly invisible to observers. The coating carries plasmons, waves that travel on the object's surface. Photons and other radiation striking a coated object cause the plasmons to ripple, channeling the energy around the object. Once the diverted energy has made its way around the coated object, it is converted back to radiation. The net result is that it seems to an observer that the radiation has traveled straight through the object. So the image of whatever is behind the object appears in front of it, making the object seem transparent.

But there's one drawback: for the coating to work, the wavelengths of the light or other radiation striking the object must be approximately the same size as the object you want to make invisible. Therefore, the "invisibility coating" only works with objects smaller than a pinhead, not large objects like human beings. So while nanobots may soon achieve invisibility, that dream for full-size people is still a long way off.

Jet Packs

HUMANS HAVE ALWAYS WANTED to fly, whether in a vehicle or under their own power. In the legend of Icarus, a young man fashions wings out of feathers and wax and gains the power of flight; but when he gets too close to the Sun, the wax melts, the wings fall apart, and he plunges to his death. The notebooks of Leonardo Da Vinci contain sketches for flying machines he hoped could be built but never were.

In science fiction, the preferred mode of personal flight is usually the jet pack: a small rocket strapped to your back, providing enough propulsion to enable you to fly without a flying car, plane, or rocket ship. The most prominent sf use of the jet pack was in Philip Nowlan's first Buck Rogers stories, "Armageddon 2419 A.D." (1928) and "The Airlords of Han" (1929).

Several science fiction movies, including *The Running Man* (1987) starring Arnold Schwarzenegger and *The Rocketeer* (1991), feature characters using jet packs. In the 1952 sf film *Radar Men from the Moon*, Commander Cody wears a helmet and rocket pack, and is clearly the model for the Rocketeer. Cody's mission is to stop the evil Retik the Ruler from conquering Earth with his lunarium ray.

It's not just science fiction writers that find the idea of "personal flying" appealing. Hang gliding enthusiasts strap what essentially is a set of wings to their back, jump off a cliff, and fly without propulsion by gliding on air currents. Hot air balloons are also a sort of "personal flying machine." In the 1960s, kits for building "gyrocopters," single-passenger helicopters, were sold by mail order through ads in comic books.

Recently, a man bought Army surplus weather balloons, tied them to a lawn chair in his back yard, and filled them with helium. Sure enough, he took off, rose thousands of feet in the air, and began to fly, pushed by air currents. His only control mechanism was a shotgun he carried while sitting in the chair: to lower his altitude when he got too cold, he shot some of the balloons.

Engineers have been striving for decades to build a practical personal jet pack for individual flying. In his book *All about Rockets and Space Flight* (1964), Harold Goodwin describes the first working jet pack:

> Wendell Moore chose hydrogen peroxide as his fuel. By controlling the flow of liquid across a catalyst of silver, he could change the thrust gradually from zero to full power and back to zero again.
>
> Two tanks of hydrogen peroxide provide thrust for the belt, each tank having its own nozzle. This propulsion unit is attached to a fiberglass frame like a corset and padded rings are attached to the frame.
>
> The belt, which weighs 110 pounds when filled with fuel, can only operate for about a half-minute. But in that time the operator can fly more than forty feet in the air, over ground distances of more than one hundred yards, and at speeds up to sixty miles an hour. Using his hand throttle he can vary the thrust of his belt from zero to three hundred pounds.

The problem Moore had, experienced by all engineers trying to devise a practical jet pack, was the limitation of weight, fuel, and distance. For a jet pack to be small and light enough for a man to carry on his back, it cannot, given the constraints of current rocket design, carry enough fuel for a flight of any significant duration or distance.

Moore's flight took place in the early 1960s. In 1969, another jet pack system was developed and tested by Bell AeroSystems. The jet pack, which weighed 126 pounds, burned kerosene for fuel, had a range of ten miles, and flew at a top speed of sixty miles per hour, which it could sustain for just about six minutes. During a test flight, pilot Robert Courter flew at an altitude of twenty-five feet. The Bell jet pack's turbine engine was mounted vertically, with the inlet down. The thrust of combustion was exhausted through nozzles pointing downward just behind the operator's shoulders. Using a pair of control arms with handgrips, the test pilot could control both the speed of the jet engine and the direction of the nozzles.

Today a California company, Millennium Jet, is working on building a practical jet pack, the SoloTrek. Weighing 325 pounds, the SoloTrek jet pack uses an internal combustion engine burning gas, kerosene, or other heavy fuel to generate lift through ducted fans. Unlike its short-flight-time predecessors, the SoloTrek can carry enough fuel for a two-hour flight. The goal is to build a SoloTrek jet pack that can fly at altitudes of up to eight thousand feet at speeds of up to eighty miles per hour.

Lasers and Ray Guns

BAD GUYS AND GOOD GUYS in science fiction sometimes shoot guns with bullets, but their preferred weapon of choice is the ray gun—a handheld weapon that shoots a beam of destructive energy that destroys the target either by melting or vaporizing whatever it hits.

Ray guns and energy guns were commonplace in 1920s and 1930s science fiction stories such as E. E. "Doc" Smith's Skylark series, starting with *The Skylark of Space* (1928). They were also prominently featured in the 1930s comic strips Buck Rogers and Flash Gordon.

In the 1960s television series *Star Trek*, the crew carries ray guns called "phasers." On the highest setting, a blast from a phaser dematerializes its target on contact. The phaser could be set to stun, kill, or vaporize the target.

The laser is the closest thing we have to a ray gun, although no handheld laser has nearly enough intensity to be used as a weapon. The word *laser* is an acronym for "light amplification through stimulated emission of radiation."

The first laser was built by Theodore Maiman in 1960. It consisted of a cylindrical ruby crystal covered with a thin silver film and held between two highly polished mirrors on either end. Energy was fed into the cylinder from a flash lamp until the ruby emitted a red light, which escaped through a tiny hole in one of the end mirrors.

Rubies are not needed to make lasers anymore. The "emission of radiation" can come from another source. In industrial lasers, gases—including helium, argon, and carbon dioxide—are commonly used to generate the laser beam.

Unlike the light emitted by a flashlight, which spreads and dissipates quickly, a laser beam remains tight and concentrated over considerable distances without spreading out. By concentrating all of the light energy on a single point, the laser beam can generate a temperature of many thousands of degrees—enough to melt through solid steel.

Today lasers have many uses. It is commonplace for eye doctors to perform surgery correcting faulty vision with lasers. You can buy a laser pointer for a few dollars at your local stationery store for use in business presentations; the laser pointer produces a dot of red light useful for pointing to portions of a slide or overhead. Industrially, lasers are used to cut metal in manufacturing operations.

Laser guns and rifles are not, however, practical weapons; but lasers are used on rifle sights to help shooters aim. The laser shines a tiny red dot on the target, and the gunner aims for that dot when he pulls the trigger and fires the bullet.

Ronald Reagan's Strategic Defense Initiative (SDI), nicknamed "Star Wars," called for a space-based defense system in which satellites, armed with lasers or other energy beam weapons, would shoot down enemy nuclear missiles launched at the U.S. To provide enough energy for the beam to destroy or deflect the nuclear missiles, each satellite would be powered by an onboard miniature nuclear reactor. The miniature nuclear reactors would be cooled by liquid metal, either lithium or sodium, since unlike an Earth-based station, large quantities of water would not be available. The beam weapons would require thousands of megawatts of electrical power to fire the short bursts of energy powerful enough to damage and destroy the enemy's weapons.

Today you can buy ray guns over the Internet. One manufacturer sells an ion ray gun that shoots a beam of charged particles. Operating on nine-volt batteries, this handheld weapon can cause a painful electric shock to the person being shot at distances of up to twenty feet.

The U.S. military is developing both microwave and laser beam ray guns. The goal is to develop a heat ray that causes an intense burning pain but does not kill the target. Such a weapon might be mounted on Humvees to control rioting crowds.

Liquid Metal

ALL METAL BECOMES LIQUID when heated to high enough temperatures, so the term "liquid metal" refers to a metal that is liquid at room temperature. Science fiction historian James Gunn comments, "As I recall, A. Merritt's *The Metal Monster* (1920) had some liquid properties, as may have Jack Williamson's first published story, 'The Metal Man' (1928). Liquid metal, as such, is often used as an image in early sf, not so often as a plot element."

Liquid metal is central to the plot in the 1991 film *Terminator 2*, the sequel to *The Terminator* starring Arnold Schwarzenegger, in which he fights an enemy more advanced than him: a Terminator made out of "liquid metal."

Because it's made out of liquid metal, this more advanced model of the Terminator can change shape as it wishes, including forming limbs into knives, axes, and other edged cutting weapons. He can also form into a liquid to move through the tiniest openings, and then resolidify into a deadly robot.

Any metal becomes liquid once it is heated to the melting point.

The melting point for most metals is in the thousands of degrees, but there are exceptions. Lead can be liquefied by heating it in a pot or pan over an open flame on your stove. Mercury, used in thermometers, is liquid at room temperature. Both are toxic, so you shouldn't experiment with either.

What about liquid metals that can change shape under some sort of electrical or other method of control? Such a class of metals, called *shape memory alloys*, does in fact exist. A shape memory alloy is a metal that contracts when an electric current heats it to a certain temperature, then returns to its original shape when the current is switched off and the metal cools. Engineers believe they may soon be able to use shape memory alloys in devices like electric seat adjustments in cars or even as muscles in robots or prosthetic devices.

A similar type of material is *magneto-rheological fluids* (MRFs). They become stiff and solid when a magnetic field is applied, and return to liquid form when the magnetic field is removed. MRFs are already used commercially in automotive suspension systems. There are also *electro-rheological fluids* (ERFs) that stiffen and solidify when electricity is applied.

Scientists recently reported the existence of a new state of matter that exhibits many of the properties of the liquid metal shown in the Terminator sequels, except that it is not a metal. This new state of matter, called a *supersolid*, has all the properties of a crystalline solid, but it flows like a liquid with zero viscosity, totally without resistance.

To make a supersolid, researchers compressed a helium isotope, helium-4, into a tiny glass disk with atom-sized pores. They then placed the glass disk into a capsule, applied pressure greater than sixty times the normal atmospheric pressure, and cooled the isotope to a temperature approaching absolute zero.

The real-life substance that most resembles the liquid metal of the Terminator sequels is *metallic glass*. In regular metals, atoms are arranged in an orderly structure, like a crystal. In metallic glass, the metal atoms are arranged in a disorderly fashion, much like the atoms in a liquid or glass. Because of this haphazard structure, metallic glass has a much lower melting temperature than conventional metals, is extremely malleable, and can be molded as easily as plastic. In addition, its non-crystalline structure eliminates the stress points inherent in crystals, making liquid glass three times stronger—and ten times springier—than industrial steel.

Men on the Moon

RITERS HAVE BEEN FASCINATED with the idea of sending men to the Moon for many centuries, hundreds of years before the technology for such a trip was developed. In the second century A.D. Lucian of Samosata wrote two satirical voyages in "Icaromenippus" and "A True Story." These were followed by Ariosto's *Orlando Furioso* (1516), Johannes Kepler's "Somnium" (1634), Francis Godwin's *The Man in the Moon* (1638) and Cyrano De Bergerac's *L'autre monde* (1657).

Edmond Rostand wrote a play, *Cyrano de Bergerac*, first performed in 1897, which was a fictional account of Cyrano's life. In the third act, the Cyrano character describes several methods he has developed for traveling to the Moon:

> I might construct a rocket, in the form of a huge locust, driven by impulses of villainous saltpeter from the rear, upwards, by leaps and bounds.
>
> Or again, smoke having a natural tendency to rise, blow in a globe enough to raise me.
>
> Finally, seated on an iron plate, to hurl a magnet in the air. The iron follows—I catch the magnet—throw again—and so proceed indefinitely.

In his 1865 novel *Moon Ship*, Jules Verne sent men to the Moon in a nine-foot-diameter capsule shaped like a bullet and shot from a nine-hundred-foot cannon. And in her 1929 story "The Girl in the Moon," Thea Von Harbou sent a woman to the Moon in a streamlined rocket with aerodynamically shaped fins. Other early "man in the Moon" stories include Daniel Defoe's *The Consolidator* (1705) and Joseph Atterly's *A Voyage to the Moon* (1827).

George Griffith's *A Honeymoon in Space* (1901) described the Moon as a dead world. On the other hand, Edgar Rice Burroughs' *The Moon Maid* (1923) depicted the Moon as capable of harboring beings that might conquer Earth. Robert Heinlein wrote frequently about the Moon in such books as *Rocket Ship Galileo* (1947), *The Man Who Sold the Moon* (1950), and *The Moon is a Harsh Mistress* (1966).

Putting a man on the Moon became a reality on July 20, 1969, when Neil Armstrong became the first man to step onto the lunar surface. Armstrong and the other eleven astronauts from the various Apollo missions are the only humans who have ever walked on the Moon.

In his book *Greetings, Carbon-Based Bipeds* (1999), Arthur C. Clarke talks about the benefits of lunar travel:

> And this leads to a major role that the Moon will play in the development of the Solar System: it is no exaggeration to say that this little world, so small and close at hand (the very first rocket to reach it took only thirty-five hours on the journey), will be the stepping-stone to all the planets.
>
> The reason for this is its low gravity; it requires twenty times as much energy to escape from the Earth as from the Moon. As a supply base for all interplanetary operations, therefore, the Moon has an enormous advantage over the Earth—assuming, of course, that we can find the kind of materials we need there. This is one of the reasons why the development of lunar technology and industry is so important.

As the Earth gets more crowded, the Moon affords us millions of acres of unused land for everything from mining to establishing a lunar colony. One company, TransOrbital, already has plans to locate a data center on the Moon. CEO Dennis Laurie comments, "The Moon is a pretty safe place to store your data...9/11 caused people to think about what data backup really means, and there is also the threat of a natural disaster here on Earth, such as a small asteroid hitting the planet."

Mind Control

TORIES ABOUT MIND CONTROL proliferated in the U.S. after Franz Anton Mesmer began practicing "animal magnetism"—called "mesmerism"—in the eighteenth century, and then, in 1842, "hypnosis." Many nineteenth-century writers based stories around mesmerism and hypnosis, including Nathaniel Hawthorne in *The Blithedale Romance* (1852) and Edgar Allen Poe in "Mesmeric Revelation" (1844).

Countless science fiction stories have been written about mind control achieved both through natural and mechanical methods. By "natural" methods, I mean the subject's thoughts and actions are manipulated using psychological, pharmacological, or psionic means. "Natural" methods of mind control include hypnosis, drugs, surgery, and electroshock treatment. In a sense, psychotherapy and prescription antidepressants can both be viewed as forms of mind control; seven percent of American children ages six to eleven take Ritalin and similar drugs to control hyperactivity. By "mechanical" methods, I mean the subject's mind is controlled or his thoughts altered using a technologically advanced machine.

For years, sf writers have experimented with their own versions of mind control. In the Telzey Amberdon series of short stories written in the 1960s by James H. Schmitz, a young girl, Telzey Amberdon, develops psionic powers as a teenager. One is telepathy, the other is mind control. She has no qualms about changing the personality and behavior of her Aunt Halet, whom she finds particularly irritating.

In Stephen King's 1980 novel *Firestarter*, Andy McGee—the father of Charlie McGee, who is the "firestarter" of the title (she can start fires with her mind)—agreed to participate in a drug experiment in college to earn extra money. He did not realize that the experiment was designed to see whether the drug could give humans psionic powers, and both he and another student, a girl (whom he later marries and who gives birth to Charlie) gain such powers. The girl develops weak telekinetic abilities. Andy develops a greater mental ability, which he calls "push":

controlling other people's actions and making them do or think as he wishes. Here is King's description of Andy McGee using his power:

"I've changed my mind," Andy said. "Take us to Albany, please."

"Where?" The driver stared at him in the rearview mirror. "Man, I can't take a fare to Albany, you out of your mind?"

Andy pulled out his wallet, which contained a single dollar bill. He thanked God that this was not one of those cabs with a bulletproof partition and no way to contact the driver except through a money slot. Open contact always made it easier to push. He had been unable to figure out if that was a psychological thing or not, and right now it was immaterial.

"I'm going to give you a five-hundred-dollar bill," Andy said quietly, "to take me and my daughter to Albany. Okay?"

"Jeee-sus, mister—"

Andy stuck the bill into the cabby's hand, and as the cabby looked down at it, Andy pushed ... and pushed hard. For a terrible second he was afraid it wasn't going to work, that there was simply nothing left, that he had scraped the bottom of the barrel when he had made the driver see the nonexistent black man in the checkered cap.

Then the feeling came—as always accompanied by that steel dagger of pain. At the same moment, his stomach seemed to take on weight and his bowels locked in sick, gripping agony. He put an unsteady hand to his face and wondered if he was going to throw up ... or die. For that one moment he wanted to die, as he always did when he overused it.

There is no doubt that drugs can alter a person's mood; this in itself could be considered a form of mind control. Sodium pentathol, for example, has been used as a "truth serum," but it's really a mild sedative. It makes people more talkative, but does not compel them to tell the truth. During World War II, CIA scientists were asked to develop a more effective truth serum, which they dubbed "truth drug" (TD). TD was nothing more than an extract from the marijuana plant. Later the CIA used LSD as a truth drug.

In 1977, the CIA allegedly conducted a series of research projects to develop an effective mind control method in which a number of subjects went insane and two people died. These programs were said to involve a combination of drugs, sensory deprivation, microwaves, psychological conditioning, surgery, and brain implants.

Both the CIA and the U.S. Navy have supposedly conducted mind control experiments. The Navy's goal was to detect and locate enemy submarines by telepathically reading the minds of their crew.

In his 1951 story "March Hare Mission," Ford McCormack imagines a mind control drug, "nepenthal," that made the subject unable to sustain thought by wiping out his or her short-term memory:

> The drug nullified one's ability to store recollections of current happenings, usually on an almost instantaneous basis. However, the presymptomatic memory was unaffected. Thus, under its influence, the subject was unaware of any lapse of time between the onset of the drug's full effect and the present moment.

Mechanical methods have also offered substantial subject matter for sf writers. In his 1954 short story "Big Game Hunt," one of several he has written on mind control themes, Arthur C. Clarke describes a mind control machine based on the technique of "neural induction":

> It has been known for many years that all the processes that take place in the mind are accompanied by the production of minute electric currents. By "playing back" the impulses he had recorded, he could compel his subjects to repeat their previous actions—whether they wanted to or not.

Michael Crichton experiments with a similar method of mind control in his novel *The Terminal Man* (1972). In the novel surgeons attempt to control seizures and violent behavior by implanting brain electrodes:

> The NPS staff has developed a computer that will monitor electrical activity of the brain, and when it sees an attack starting, will transmit a shock to the correct brain area. This computer is about the size of a postage stamp and weighs a tenth of an ounce. It will be implanted beneath the skin of the patient's neck.
>
> We will power the computer with a Handler PP-J plutonium power pack, which will be implanted beneath the skin of the shoulder. This makes the patient completely self-sufficient. The power pack supplies energy continuously and reliably for twenty years.

In another 1954 story, "Patent Pending," Clarke describes a technology for manipulating the mind by recording memory and thoughts for later use:

It was simply the natural extension of what man has been doing for the last hundred years. First the camera gave us the power to capture scenes. Then Edison invented the phonograph, and sound was mastered. Today, in the talking film, we have a kind of mechanical memory which would be inconceivable to our forefathers.

But surely the matter cannot rest there. Eventually science must be able to catch and store thoughts and sensations themselves, and feed them back into the mind so that, whenever it wishes, it can repeat any experience in life, down to its minutest detail.

It was done electronically, of course. You all know how the encephalograph can record the minute electrical impulses in the living brain—the so-called "brain waves" as the popular press calls them. Julian's device was a much subtler elaboration of this well-known instrument. And, having recorded cerebral impulses, he could play them back again. It sounds simple doesn't it? So was the phonograph, but it took the genius Edison to think of it.

In his 1972 short story, "I Tell You, It's True," Poul Anderson describes another such mind control machine:

Work on synergistics had suggested that the right combination of impulses might trigger autocatalytic transformations in the synapses. It doesn't take a lot, you see. These events happen on the molecular level. What's needed is not quantity but quality: the exact frequencies, amplitudes, phases, and sequences.

Nothing happens except that the subject believes absolutely what he's told or what he reads while he's in the inducer field. There doesn't seem to be decay of the new pattern afterward. Why should there be? What we have is nothing but an instant reeducator.

Mechanical methods of mind control have also been used in the real world. According to a 2002 article in *The Economist*, a team of neuroscientists, investigating a treatment for depression, used electrodes to stimulate the brains of female patients in ways that caused pleasurable feelings.

There are persistent rumors that, during the Cold War, both the Soviets and the United States ran covert programs to identify and develop men and women with psi powers for espionage and defense. A telepath, for instance, could read the minds of enemy generals and discover their military plans. An agent with psi powers could prevent an enemy soldier in a missile silo from firing the weapon when ordered.

Some sources claim that individuals with psionic powers can alter computer files with their mental powers. If that's true, it's a short step to being able to control computers and computer-based systems, such as weapons controls.

There is no doubt that electromagnetic radiation can affect human biology. Generators producing electromagnetic fields are routinely implanted in patients with spinal injuries to promote accelerated bone growth and repair of the spine.

In his book *Mind Control, World Control* (1997), Jim Keith writes that researchers working for the CIA bombarded the brains of lobotomized monkeys with radio waves. He also claims that the U.S. military has experimented with extremely low frequency electromagnetic transmissions (ELFs) designed to "cause aberrations in the thought processes of human beings" including hallucinations, disordered thought, confusion, depression, anger, and hopelessness.

In addition, writes Keith, the CIA allegedly has developed a mind control technology called Radio Hypnotic Intracerebral Control Electronic Dissolution of Memory (RHIC-EDOM). Induced through electromagnetic broadcasting or implantable devices, the RHIC-EDOM system can remotely induce a hypnotic state and impart hypnotic commands without the subject knowing this was done.

Another mind control system, "acoustic psycho-correction," supposedly developed by the Russians, transmits specific commands via static or white noise bands into the human subconscious. The low-frequency transmission reaches the brain not through the ear but through bone conduction, so ear plugs cannot block the mind control signal.

Possession is also a form of mind control, and Frederik Pohl explores this concept in his 1984 novel *Demon in the Skull*:

> Then came the possessors. The witches, the aliens, the whatever-they-were that smote the Earth that Christmas season.
>
> The word became current at once: they were "possessed."
>
> Possessed by what? No one could say. These were the first cases of "possession" that the world had seen—or at least acknowledged—since the great casting out of devils of the Middle Ages, and the world had long since abandoned its belief in such things.

The belief did not stay abandoned...especially when, once the large-scale events had taken place, slowly and increasingly began the terror-reign of small ones. Of rapes and murders and suicides and arsons, every one of them committed by some person who, if he survived, wound up weeping and stammering in terror and pleading for understanding that he had not been able to help himself.

Alien mind control of a human is a popular sf theme. Steven Spielberg's TV miniseries *Taken* portrays the aliens who crash landed at Roswell as able to exercise mind control over humans. Alien mind control is also central to the plot of Stephen King's short story, "I Am the Doorway" (1971):

And little by little, I felt them. Them. An anonymous intelligence. I never really wondered what they looked like or where they had come from. It was moot. I was their doorway, and their window on the world. I got enough feedback from them to feel their revulsion and horror, to know that our world was very different from theirs. Enough feedback to feel their blind hate. But still they watched. Their flesh was embedded in my own. I began to realize that they were using me, actually manipulating me.

In this story, an astronaut returning from Venus is "infected" by an alien who begins taking over his body. The alien controls his actions and begins to alter his body, making the fingers shrink and growing eyes on the palms of his hands, which he destroys by soaking his hands on kerosene and lighting them on fire. But he is unable to triumph; the eyes appear on his chest, and the story ends with him preparing to kill himself with a shotgun.

Mutations

HUGO DE VRIES popularized the concept of "mutation" in biological evolution in *Die Mutations-Theorie* (1901–3). In 1927, H. J. Muller induced mutations in fruit flies by irradiation.

As for mutants in science fiction, John Taine was the first to use this idea in *The Greatest Adventure* (1929), *The Iron Star* (1930), and *Seeds of Life* (1931). Since then mutations have become a standard element in pulp stories, comics, serious science fiction, and film, including monsters and superheroes.

Among the dozens of science fiction novels and stories featuring mutants:

- Jack Williamson's "The Metal Man" (1928).
- Edmond Hamilton's "The Man Who Evolved" (1931).
- H. G. Wells' *Star Begotten* (1937).
- A. E. van Vogt's *Slan* (1940).
- Robert A. Heinlein's "Universe" (1941).
- Henry Kuttner's "I Am Eden" (1946).
- Wilmar Shiras' *Children of the Atom* (1948).
- F. N. Waldrop's *Tomorrow's Children* (1947).
- Judith Merril's "That Only a Mother" (1948).
- Fritz Leiber's "Coming Attraction" (1950).
- Isaac Asimov's *Foundation and Empire* (1952).
- Edgar Pangborn's *Davy* (1964).
- John Wyndham's *The Chrysalids* (1955).
- Walter M. Miller's *A Canticle for Leibowitz* (1955).
- Harlan Ellison's "The Discarded" (1959).
- Lester del Rey's *The Eleventh Commandment* (1962).
- Samuel R. Delany's *The Einstein Intersection* (1967).
- Norman Spinrad's *The Iron Dream* (1972).
- Sterling Lanier's *Hiero's Journey* (1973).

- A. A. Attanasio's *Radix* (1981).
- Greg Bear's *Blood Music* (1985).

"Mutation" and "science fiction" are closely linked in many people's minds; and they think of mutation as a strictly science fictional concept. In reality, mutation is a common natural phenomenon that plays a key role in human, animal, and plant evolution.

Charles Darwin knew that natural selection—the breeding of physiological improvements into animals and plants—was responsible for the evolution of species. But he was not completely sure what caused those improvements to happen in the first place. As he wrote in *The Origin of Species* (1859): "Whatever the cause may be of each slight difference between the offspring and their parents—and a cause for each must exist—we have reason to believe that it is the steady accumulation of beneficial differences which has given rise to all the more important modifications of structure in relation to the habits of each species."

The "cause" Darwin was searching for is mutation, a natural and normal process, and not—as science fiction often portrays it—an aberration (except rarely). As Julian Huxley explains in his book *Evolution in Action* (1953):

Mutation merely provides the raw material of evolution; it is a random affair, and takes place in all directions.

Genes are giant molecules, and their mutations are the result of slight alterations in their structure.

Some of these alterations are truly chance rearrangements, as uncaused or at least as unpredictable as the jumping of an electron from one orbit to another inside an atom.

Others are the result of the impact of some external agency, like X-rays, or ultra-violet radiations, or mustard gas. But in all cases they are random in relation to evolution.

To put the matter in a nutshell: the capacity of living substance for reproduction is the expansive driving force of evolution; mutation provides its raw material; but natural selection determines its direction.

One would expect that any interference with such a complicated piece of chemical machinery as the genetic constitution would result in damage. And in fact this is so; the great majority of mutant genes are harmful in their effects on the organism.

A few have been favorable in combination with other genes. Selec-

tion automatically incorporates this tiny minority of favorable varia-
tions in the hereditary constitution, by shifting them from the mass
of unusable dross.

Radiation—whether exposure to nuclear radiation or excess exposure to the ultraviolet rays of the Sun—often causes mutation. Isaac
Asimov and Linus Pauling also postulated that decay of carbon-14 in
our own bodies can trigger mutations. Adrian Berry, in his 1994 *The
Book of Scientific Anecdotes*, quotes Asimov's explanation:

> The key molecules of the cells are its genes. It is change in the genes
> which can bring about mutations and cancer.
> But now let's consider carbon-14. The gene is not merely being
> shot at by beta particles from exploding carbon-14 atoms. It *contains*
> carbon-14 atoms—on the average, one carbon-14 atom in the genes
> of every two cells. Each second, in your body as a whole, fifty car-
> bon-14 atoms located in the genes are exploding and sending out a
> beta particle. Even if you suppose that the beta particle might plow
> through the gene without hitting any of its atoms squarely enough to
> do damage, the fact still remains that the exploded carbon-14 atom
> has been converted to a stable nitrogen-14 atom. By this change of
> carbon to nitrogen, the gene is chemically altered.
> Furthermore, the carbon-14 atom, having shot out a beta particle,
> recoils just as a rifle would when it shoots out a bullet. This recoil
> may break it away from its surrounding atoms in the molecule, and
> this introduces another change. Carbon-14 is much more likely to be
> responsible for "spontaneous" cancer and mutations than potassium-
> 40 is.

But the mutations Huxley describes are small or incremental, meaning evolution is slow and gradual—not greatly visible to the untrained
observer. They are also viable, meaning they can survive, and individuals containing the mutant gene can procreate, propagating the mutation.

The type of mutation featured in science fiction is usually dramatic,
instant, aberrant, and nonviable; the creature with the mutation does
not long survive. Example: Paul Blaisdell's horned monstrosity in the
1955 film *The Day the World Ended*.

Pollution and radiation have caused mutations that seem more like
science fiction than science fact. For instance, pesticides and other pol-

lutants have created thousands of deformed frogs throughout North America. Researchers at McMaster University in Ontario, Canada, found that the lab mice exposed to the exhaust fumes from two steel mills and a busy highway were twice as likely to give birth to mutated offspring as mice breathing filtered air. The scientists believe that the mutations are caused by microscopic airborne particles of soot and dust containing toxins such as polycyclic aromatic hydrocarbons.

Excessive exposure to the ultraviolet radiation of the Sun's rays can also cause genetic mutation. As the ozone layer has been thinning as the result of chlorofluorocarbons and other chemical pollutants, more ultraviolet radiation reaches the planet's surface, potentially causing more mutations. Frogs are particularly susceptible. Because they lay eggs without shells and have permeable skin, frogs and other amphibians are particularly susceptible to pollutants, radiation, and other factors that can cause mutation. Therefore, mutated frogs are an "early warning" sign that mutations in other, less susceptible species may follow. The most common deformities are missing legs, missing toes, deformed legs, and extra legs. One frog found in the Arctic National Wildlife Refuge had its ankle connected to its pelvis by a thin thread of tissue. Other frog deformities reported include lack of eyes and eyes coming out of the stomach.

Since 1995, deformities have been found in sixty species including frogs, salamanders, and toads.

Nanotechnology

HE "NEXT BIG THING" in technology is actually very small.

Nanotechnology gets its name from the nanometer, which is one-billionth of a meter—1/75,000 the size of a human hair. A nanometer comprises about four individual atoms.

Here's why that matters so greatly: the classical laws of physics change when dealing with matter that small. Once scientists can manipulate atoms—which combine to form molecules, the building blocks of our natural world—they can use the altered laws of physics to create new building blocks that produce new materials with the exact properties they desire: smaller, stronger, tougher, lighter, and more resilient than what has come before.

Michael Crichton's *Prey* (2000) is probably the first best-seller based entirely on a nanotechnology theme. A colony of "nanobots"—molecule-sized robots—develop artificial intelligence and run amok. Writing in *Parade* magazine, Crichton suggests that a mass of tiny nanobots, smaller than specks of dust, could be programmed to travel in a cloud over an enemy nation and send back reconnaissance photos. Such a swarm of nanobots could not be shot down; bullets would pass right through the cloud.

In the 2001 movie *Jason X*, supernatural killer Jason Voorhees of Friday the 13th fame is defeated by an android, but nanotechnology revives him as an unstoppable cyborg, and he wins the rematch.

In his 1956 short story "The Next Tenants" Arthur C. Clarke describes tiny machines that operate on a micro (thousandth of a meter) scale, not the nano (billionth of a meter) scale, but the basic idea is similar:

> It's a micromanipulator. These were devices with which, by the use of suitable reduction gearing, one could carry out the most incredibly delicate operations. You moved your finger an inch—and the tool you were controlling moved a thousandth of an inch. The French scien-

tists who had developed this technique had built tiny forges on which they could construct minute scalpels and tweezers from fused glass. Working entirely through microscopes, they had been able to dissect individual cells.

In his 1969 short story "How It Was when the Past Went Away," Robert Silverberg describes nanodevices used as components of a stereo loudspeaker:

> He began to paint the inner strips of loudspeakers on: a thousand speakers to the inch, no more than a few molecules thick, from which the sounds...would issue in resonant fullness.

Richard Feynman, one of the physicists who helped develop the atomic bomb, was probably the first in the science community to conceive of the idea of nanotechnology. In 1959 he observed that the laws of physics did not prohibit machines, including motors and computers, from being built, one atom at a time, at the molecular level. K. Eric Drexler introduced the concept of nanotechnology to the general public and popularized it in *Engines of Creation* (1987). It was then seized upon by sf writers as a kind of catch-all technology that enables characters to do or be almost anything, making possible cures for disease and even of aging. Ben Bova wrote about nanotechnology in *Voyagers III: Star Brothers* (1990), Greg Bear in *Queen of Angels* (1990), and Linda Nagata in *Limit of Vision* (2002). In Neal Stephenson's *The Diamond Age* (2000), visitors to cities are identified by airborne nanosensors, and books are made of nanoparticles that can modify the text to suit reader preference.

Feynman suggested that the manufacturing of nanomachines would require the ability to manipulate individual atoms. He also said we would need to increase the magnifying power of electronic microscopes one hundredfold, so that we would be able to see individual atoms. Today, the atomic force microscope (AFM) can accomplish both of those tasks. Not only does an AFM allow the user to see atoms, but he or she can manipulate them, too. The AFM uses a tiny needle, composed of a single atom, to read the surface of a compound or construct at the nanoscale directly, much like a turntable stylus running over the surface of a vinyl record.

Nanotechnology could lead to incredible advances, some merely practical, others almost sublime. On the most mundane level: scratch-proof glass and tiles that shed dirt and never need cleaning. On a more sophisticated level: precision drug-delivery systems...super-fast computers the size of a sugar cube...desktop storage drives that could hold the entire Library of Congress...and building blocks to produce new materials with the exact properties desired. A machine built on the nanoscale might be one hundred nanometers long: about one thousand times smaller than the diameter of a human hair.

Just look at what nanotech researchers have already developed:

- A new miniaturized, nanotechnology-based drug dispenser for diabetics, implanted just under the skin of the arm, releases controlled dosages of insulin for up to six months—eliminating the need for patients to give themselves daily injections.
- Carbon nanotubes have been used to build a tiny, implantable detector that could one day allow diabetics to continuously monitor their glucose levels, eliminating the need to do daily blood tests. At Rice University, nanoparticles known as "nanoshells" have been used as chemical detectors with ten thousand times greater sensitivity than traditional methods.
- Scientists in nanotechnology research facilities have arranged individual carbon atoms in microscopic tubes that are one hundred times stronger than steel but much lighter. And researchers at Purdue University have made nanotubes from DNA. The properties of the DNA-based nanotubes can be controlled by adding different molecules to their structure. For instance, attaching nylon molecules to the DNA nanotube's surface could make long, flexible nylon fibers that could be used in body armor and parachutes.
- By applying a small electrical current to a carbon nanotube, scientists at the Lawrence Berkeley National Laboratory have created a nanoscale conveyor belt capable of carrying individual atoms, in much the same fashion that an ordinary conveyor belt on an automobile assembly line carries car parts. The ability to deliver individual molecules to specific locations will aid in the fabrication of nanoscale devices. In addition, Fraser Stoddart, a chemist at UCLA, has built a molecular piston—driven by chemical reactions, light, or electricity—that could be used to drive nanoscale machines.
- At Tel Aviv University, scientists are using peptide molecules— short chains of amino acids—to fabricate silver nanowires. The peptide molecules, which naturally form tiny tubes, are placed in

solution with silver ions, producing silver-filled nanotubes. An enzyme is used to dissolve the outer shell of peptide, leaving silver nanowires that can be used in microelectronic devices.

- Scientists at Hewlett-Packard have potentially found a way to replace silicon transistors with "nanoswitches" consisting of electrically switchable molecules. The key component of their nanoswitch is a crossbar latch measuring just a single layer of molecules thick. The latch can be flipped to binary 0 or 1 and back again, and preserve the output of that computation for use in subsequent calculations.

- Dr. Charles Lieber, a researcher at Harvard University, recently completed a nanotechnology-based sensor capable of detecting a single molecule of a given substance. Using this sensor, Dr. Lieber demonstrated the ability to test for a protein known as the prostate-specific antigen, or PSA, a fairly reliable marker for prostate cancer. The nanosensor has almost ten times greater sensitivity than any existing PSA test, which can mean early detection of prostate cancer—and greater survival rates for patients.

- A molecule known as vascular endothelial growth factor (VEGF) instructs cells on the inner linings of blood vessels to replicate. Researchers at the Mayo Clinic have used gold nanoparticles to block VEGF's ability to stimulate blood vessel cell growth. By cutting off blood supply to tumors, the gold nanoparticles might be able to help treat cancer.

- Researchers at Northwestern University are using bacteria, viruses, fungi, and other microbes to build materials on the nanoscale. In one experiment, nanoscopic gold particles were attached to short DNA strands in a dish of nutrient-containing fungal spores. As the fungus grows, it forms tubes, and the gold particles become attached to the tubes. When the tubes are dried and pressed, they form a fibrous golden material.

- At NASA's Ames Research Center, genetic engineering is being used to create new nanostructures. The researchers removed a gene from a single-celled bacterium that lives in geothermal hot springs and modified it to produce large quantities of a heat-resistant protein. The protein is assembled in ring structures and crystallized into wafers used in the microelectronics industry.

- Nanoscale devices have the potential to be used as highly accurate sensors. A nanoscale sensor could be built that is not much bigger than a single molecule of the substance it is designed to detect, making it especially accurate and sensitive. Researchers at Rice University have shown that the electromagnetic field around nanoparticles can be controlled, which could lead to the development of a nanodevice capable of analyzing samples as small as one molecule.

- One company is using nanotechnology to make self-cleaning window glass; another is making a nano-based wound dressing with antibiotic and anti-inflammatory properties.
- Hydroxyapatite (HPA) is a calcium phosphate material commonly used in prosthetic hips, discs, and bones. Researchers are using nanoparticles of HPA to develop new artificial bone implants that behave exactly like or better than real bone, mimicking its strength and porosity while being accepted by the body's immune system.
- At the University of Notre Dame, scientists have developed single-wall carbon nanotubes with ferrocene molecules attached to the tube walls. If the carbon nanotubes are exposed to solar radiation, electrons flow from the ferrocene to the carbon atoms—the principle on which photovoltaic solar cells operate—a first step in developing nanotechnology-based solar power.
- Scientists at Virginia Commonwealth University have created a nanofiber mat that could be placed directly on a cut to initiate blood clotting. The nanofibers have a diameter of ten nanometers, approximately the size of a skin cell. They mimic the body's natural tissues and are absorbed as the wound heals.
- Lasers are being used to manipulate nanoparticles. A highly focused laser beam traps and moves objects ranging in size from a protein (five nanometers) to a cluster of a dozen cells (one hundred nanometers). In one application lasers are used to manipulate particles within colloidal crystals, deliberately creating defects. The altered crystals were able to amplify and switch signals in optical telecommunications networks. The crystals may soon be used in the manufacture of optical sensors for homeland security.
- Another nanotechnology innovation developed by IBM is Millipede, a hard drive built on the nanoscale. A computer memory device, the nanodrive is about the size of a postage stamp, with a data storage capacity of several gigabytes. Much like the atomic force microscope, the nanodrive operates like a phonograph

needle scratching the surface of a vinyl record. The sharp tips of miniature silicon cantilevers read data inscribed on a polymer medium. To read the data, the tips are heated to about 570 degrees Fahrenheit. When the silicon tip encounters a groove or depression in the polymer's surface, it transfers heat to the plastic, and the silicon "stylus" cools by a few thousandths of a degree. A digital signal processor converts temperature fluctuations in the silicon stylus into a data stream.

And that's just the tip of the iceberg when it comes to the benefits nanotechnology is going to bring the world in fields ranging from materials science and pharmaceuticals, to semiconductor manufacturing and computer storage, to construction and consumer goods.

The U.S. Government is bullish on the nanotechnology industry. In December 2003, the federal government passed a Nanotechnology Research and Development Act allocating $3.7 billion for nanotech research and development from 2005 to 2008. The National Science Foundation estimates that the global market for nanotechnology-related products will have reached $1 trillion by 2005.

The science is not without its problems, however. One concern is that self-replicating nanoscale machines and nanorobots will not only create duplicates of themselves, but that they may evolve into something other than their original form. In the new form, the evolved software may have goals conflicting with our own. If a nanorobot replicates and evolves quickly, and turns destructive toward its creators, what hope would we have? They would be likes bugs, bacteria, or viruses, in that they could infect, infiltrate, and damage us with ease. We, in turn, could do little against them; no matter how many we destroy, more will be built to take their place. As Goliath discovered when battling David, sometimes the most dangerous enemy is not the largest, but the smallest.

Is self-replicating nanotechnology just science fiction? No. It has already become science fact. IBM researchers are hot on the trail of developing computer chips that put themselves together. They have already demonstrated a computer chip that can assemble a portion of itself on its own. In the finished chip, nanocrystalline cylinders act as capacitors where data is stored.

Neutron Stars

I N 1916, astronomers first proposed the theory that gravity could cause older stars to collapse into small, dense balls of matter.

For a star approximately one-and-a-half to three times the mass of our Sun, the nuclear reactions in the core go out of control as that region collapses. This collapse causes an explosion, and the star blows most of its mass out into space—this is what is known as a supernova.

What remains of the star's matter shrinks down into a small, dense body. Protons collide with electrons to form neutrons, leaving a superdense star whose core is composed of solid neutrons—a neutron star. These stars have densities as high as ten million tons per cubic centimeter. This is about nine trillion times the density of ordinary matter.

Pulsars, discovered in 1968, may be emissions from rapidly rotating neutron stars. One may lie at the heart of the Crab Nebula.

The first science fiction story to feature a neutron star is Larry Niven's "Neutron Star," published in *Worlds of If*, October 1966. Here's how Niven described the birth of a neutron star:

> In such a mass the electron pressure alone would not be able to hold the electrons back from the nuclei. Electrons would be forced against protons to make neutrons. In one blazing explosion most of the star would change from a compressed mass of degenerate matter to a closely packed lump of neutrons: neutronium, theoretically the densest matter in this universe. Most of the remaining normal and degenerate matter would be blown away by the liberated heat.

In Arthur C. Clarke's 1970 story "Neutron Tide" a spaceship is destroyed when the intense gravity of a neutron star pulls it apart:

> Flatbush ran straight into the gravity well of a neutron star—a sphere of ultimately condensed mater, only ten miles across, yet as massive

as a sun—and hence with a surface gravity one hundred billion times that of Earth.

Robert Forward's 1980 novel *Dragon's Egg* deals with a race of aliens, the Cheela, that live on a neutron star, despite the crushing gravity—sixty-seven billion times greater than Earth's. Gregory Benford wrote about neutron stars in *The Stars in Shroud* (1978).

Nuclear Energy

FUSION REACTIONS have made their way into science fiction in two main ways: nuclear war (see "Nuclear War") and nuclear energy.

Pessimistic scientists and science fiction writers imagine a nuclear war that destroys most of the world's population. Often these stories don't specify the type of atomic weapon being used, but it would likely be a fusion bomb (also known as a "hydrogen bomb"), which has many times the blast force of the fission bomb (also known as an "atom bomb," used in World War II on Hiroshima and Nagasaki). Examples include Peter Bryant's *Red Alert* (1958) and Eugene Burdick's and Harvey Wheeler's *Fail Safe* (1962). James Gunn writes about nuclear fusion as an energy source in "Child of the Sun" (1977).

So, how do you take the same reaction that powers the Sun and the hydrogen bomb, and turn it into a safe, reliable source of electricity?

In his book *The Neutron Story* (1959), physicist Donald Hughes explains how fusion works:

> The process of fusion is one in which very light positively charged nuclei, such as hydrogen and lithium, combine to release energy.
>
> The great obstacle in the path to fusion arises because the light nuclei, unlike neutrons, are electrically charged, and the repulsion makes it very difficult to get them to join together, or to fuse.
>
> The only way to cause nuclei to fuse is to set them into rapid motion so that two can touch in spite of the repulsive electrical force. And to get the atoms of a material into rapid motion means raising them to an extremely high temperature—to many millions of degrees, in fact. Only at such an enormous temperature are the nuclei moving fast enough to combine and release energy by fusion.

At this temperature, the electrons are stripped from the nuclei, resulting in a super-hot gas, called plasma, consisting of ions.

The reason a practical fusion reactor has not yet been built is that scientists have been unsuccessful in finding a way to contain plasma for an extended period. The plasma can't be put in a metal or glass vessel because the heat will melt the walls. The only likely solution is holding the plasma within a magnetic field.

How would a fusion reactor work? Were it to be located in a power plant, the most practical choice seems to be the fusing of deuterium and tritium. But there are a number of problems. First, while deuterium is abundant—you can extract it from seawater—tritium is not. But you can "breed" (create) tritium in a nuclear reactor by bombarding lithium with neutrons. The fusion of deuterium and tritium also produces more tritium. Second, starting the deuterium/tritium fusion reaction requires a temperature of approximately forty million degrees—a tremendous amount of energy. To generate that kind of heat in a fusion bomb, a fission bomb is used as the detonator. Researches at the Laboratory for Laser Energetics at the University of Rochester have also experimented with using a high-powered laser beam to heat plasma to the temperature at which fusion can take place. Third, when deuterium and tritium are heated to this temperature, they become plasma, which is difficult to contain and control. A magnetic field would be needed to keep the plasma from melting the reactor walls.

When the deuterium and tritium are heated, the electrons are stripped away, leaving helium nuclei. These have a positive charge and move energetically. As fresh fuel is injected, its cold deuterium and tritium atoms collide with the heated helium nuclei already in the reactor, and are heated in turn. The deuterium and tritium fuse, giving off a neutron in the process. The emitted neutrons move in a straight line through the reactor walls with little loss of energy. A lithium blanket that surrounds the fusion chamber absorbs the neutrons and their energy. As the neutrons bombard the lithium, they generate heat. A heat exchanger transfers the heat to a conventional steam plant where it generates electricity. At the same time, the neutron bombardment of the lithium breeds tritium, which is separated and fed back to the reactor as fresh fuel.

A small-scale fusion reactor has been built and successfully tested at Princeton University. Known as the Tokamak Fusion Test Reactor, it uses a doughnut-shaped magnetic field to hold the plasma. By changing the magnetic fields using induction coils that exceed a million amperes and bombarding the fuel with neutral atoms, the plasma has been heated to a temperature of four hundred million degrees Kelvin, enabling deuterium/tritium fusion to take place. This generated 5.6 million watts of power. Unfortunately, it took more than 5.6 million watts of power to heat the plasma to this level, so the Tokamak reactor does not generate a positive net power output. In other words, it takes more energy to run the fusion reactor than the reactor can produce! Still, it is a first step toward making a controlled nuclear fusion reaction a viable alternate energy source.

When it comes to nuclear energy, however, someone beat both scientists and science fiction writers to the punch: Mother Nature. A deposit of uranium-235 in rocks in Gabon, West Africa, was so high that it reached critical mass—and triggered a sustained nuclear reaction (fission, not fusion) reaction—nearly two billion years ago. Evidence suggests that this "natural" nuclear plant generated about one hundred kilowatts of power, remaining active for 150,000 years before fizzling out. Groundwater in the rocks serves as a natural coolant, controlling the chain reaction to prevent an explosion.

A concentration of volatile xenon isotopes found in the region suggests that the reactor constantly cycled on and off, much like the geyser Old Faithful in Yellowstone National Park. For about half an hour, the fission reaction would heat the rocks, turning the water to steam, slowing the reactions down. Then the reactor would lay dormant for about two-and-one-half hours; the rock would cool and the water would seep in again, speeding up the reaction.

Geophysicist Marvin Henderson believes there is natural nuclear fission still going on underground today, within the Earth's core. Uranium is a heavy element, so when Earth was young and molten, says Henderson, most of its uranium would have sunk to the center of the planet—enough to ignite a fission reaction generating the intense heat that currently exists in the inner core.

Nuclear War (see also "Atomic Warfare")

I F AN UNLIMITED POWER SOURCE that produces its own fuel is the optimist's view of nuclear fusion, then nuclear war is surely the pessimist's. And in this regard, sf writers seem to be more pessimistic than positive.

In their novel *Deus Irae* (1976), Philip K. Dick and Roger Zelazny imagine a post-nuclear-holocaust future in which the man who actually threw the switch that started the atomic war is worshipped as a god. And in the Ray Bradbury short story "The Highway," published as part of his 1951 book *The Illustrated Man*, a farmer in the country discovers that nuclear war has begun when he observes the exodus of thousands of cars driving past his house on their way out of the city.

The original Planet of the Apes series of movies also envisioned a future Earth, decimated by nuclear war, cohabitated by two enemies: intelligent apes and a band of the remaining humans, horribly mutated, living underground and worshipping the last active nuclear bomb.

In his 1958 short story "Daybroke," Robert Bloch describes global nuclear war as observed by a sheltered survivor:

> Up in the sky the warheads whirled, and the thunder of their passing shook the mountain.
>
> All night long the mountain trembled, and the seated man trembled too; not with anticipation but with realization. He had expected this, of course, and that was why he was here.

According to a study from the World Health Organization, a nuclear war between the U.S. and Russia could kill two billion people, one-third of the Earth's population. The survivors would be left in a world devastated by the destructive force unleashed by the thousands of nuclear warheads triggered during the conflict by both sides. The shock waves created when the bombs explode could knock down buildings and flatten the surrounding area for miles. The area immediately surrounding ground zero would have radiation levels high enough to kill millions human beings farly quickly. Millions more would develop cancer and die a few years later. In addition, the wind would sweep up this radiation into the atmosphere, where it could be distributed to more rural areas thousands of miles away.

The nuclear explosions would blast huge quantities of soil and particles into the atmosphere, creating more than one hundred thousand tons of fine, dense radioactive dust for every megaton exploded. This thick cloud of dust would cover the Earth, blocking out much of the sunlight. The result is "nuclear winter": a prolonged period of cold weather and perpetual darkness. For weeks afterward and possibly longer, the global temperature could drop by as much as forty degrees Fahrenheit. In addition, the exploding nuclear devices would generate temperatures at the point of detonation in excess of three thousand to four thousand degrees Centigrade. The heat would ignite large, out-of-control fires both in the cities and the forests, releasing several hundred million tons of smoke and soot that would increase the duration and severity of the nuclear winter created by the particles and soil already kicked into the air by the blasts. The dark and cold of nuclear winter would kill vast quantities of crops and other vegetation on which humans and animals depend for sustenance. Global food shortages and high levels of radiation from the fallout could result in millions or even billions of additional deaths, leaving only a tiny population of survivors.

What's more, nuclear explosions would produce large amounts of nitrogen oxides (NOx), a pollutant known to deplete the ozone layer in the Earth's stratosphere. The ozone layer shields us from much of the Sun's ultraviolet rays, known to cause both skin cancer and mutation. The massive quantities of NOx compounds released into the atmosphere by nuclear war could wipe out a large portion of the ozone layer. When the dust and dirt causing nuclear winter finally settled, ultraviolet rays from the Sun would bombard the Earth, increasing the incidence of skin cancer.

Scenes of nuclear winter are shown in *The Matrix* (1999): the sky is always filled with dark clouds and there is almost no sunlight. In one of the only joyous moments of the series, Trinity and Neo take a ship to the surface, and for a brief few seconds break above the ever-present barrier of black clouds and perpetual lightning storms. Trinity is awestruck by a sight no other human on Earth has seen for hundreds of years: a clear blue sky and the brightly shining Sun. Neo unfortunately cannot see it, having been blinded in a fight with a man possessed by Agent Smith.

In the DC Comics' series *Superman: At Earth's End*, an aging Superman on a post-nuclear-war Earth gradually loses his powers because solar energy is unable to reach him. When taken aboard a hover craft piloted by androids, it takes a full year under the Sun to restore most of his powers.

The possibility of catastrophic global nuclear war is not merely a fear imagined by sf writers; it is an ever-present threat that's become a part of modern existence. The existing stockpile of nuclear weapons could destroy the world many times over, and all it takes is one bad decision from a president or military leader to set it off.

Parallel Universes and Parallel Worlds (see also "Alternate Universes," "Other Dimensions")

WHAT IS THE DIFFERENCE between a parallel universe and an alternate dimension?

An *alternate dimension* (see "Other Dimensions") is an area of time and space completely inaccessible from our own.

A *parallel universe* is a duplicate of our universe (which may be either a perfect or imperfect copy) that we *can* (theoretically) reach—because it exists in our own dimension, but at a great distance in space. In a parallel universe everything in our universe—galaxies, stars, solar systems, planets, even people—is duplicated. A "parallel world" implies a planet that is a duplicate of Earth.

In an 2003 article in *Scientific American*, cosmologist Max Tegmark says that the most popular cosmological model today predicts that the Milky Way has a twin galaxy, containing a twin of Earth—which in turn contains a twin of you—at a distance of 10 to the 10^{28} meters from us.

Writes Tegmark, "In infinite space, even the most unlikely events must take place somewhere." Our own universe is simply one small part of a larger "multiverse" which, somewhere, contains another universe that is a duplicate of our own.

The basic idea is this: If you arrange all of the atoms in a universe randomly in different ways, and do that an infinite number of times, probability says that you will, at some point, duplicate exactly—down to the very last blade of grass—the pattern of atoms in our own universe. Since the universe is infinite, there are an infinite number of collections of atoms in various locations in the cosmos. Probability and statistics say that one of these collections, eventually, somewhere, through random arrangement, has to be an exact duplicate of our own—a parallel world.

Another type of parallel world or universe—explored only in science fiction—is a double of our own, whether identical or imperfect, created by humans with psionic powers: the person literally creates the parallel universe with his or her mind.

An interesting twist on this angle is Alfred Bester's 1953 short story "Disappearing Act," in which future soldiers, tired of fighting an endless war, develop the psionic ability to travel backward in time to escape from the present. But there's a catch: since the soldiers' knowledge of history is imperfect, they travel backward to a period of history *as they imagine it*, not as it actually was—literally creating a parallel world as they do so.

Perhaps the earliest science fiction stories featuring a parallel universe are "The Plattner Story" and "The Remarkable Case of Davidson's Eyes," both written by H. G. Wells in the 1890s. In his 1976 novel *A Wreath of Stars*, Bob Shaw describes two parallel worlds, each made up of a different type of matter, coexisting in the same universe.

In Philip José Farmer's 1981 novel *The Unreasoning Mask*, multiple universes exist. Each is contained within a different cell of the body of God, a growing being still in the infant stage. The only way to reach alternate universes is to travel through the cell walls, which is harmful to the God body.

And in Farmer's World of Tiers series, a group of immortal Lords live in their own private, artificial universes, each with unique physical laws and characteristics. The universes were created by the Lords' ancestors, the Lords themselves no longer having access to the technology that created them. As Farmer explains in his novel *The Lavalite World* (1977),

> He still couldn't believe that Callister, whose real name was Urthona, and Red Orc and Anana were thousands of years old. Nor that they had come from another world, what Kickaha called a pocket universe. That is, an artificial continuum, what the science fiction movies called the fourth dimension, something like that.

The Lords, as they called themselves, claimed to have made Earth. Not only that, the Sun, the other planets, the stars—which weren't really stars, they just looked like they were—the whole damn universe. In fact, they claimed to have created the ancestors of all Earth people in laboratories.

Not only that—it made his brain bob up and down, like a cork on an ocean wave—there were many artificial pocket universes. They'd been constructed to have different physical laws than those in Earth's universe.

Apparently, some ten thousand or so years ago, the Lords had split. Each had gone off to his or her own little world to rule it. And they'd become enemies...

The discovery of string theory, which posits eleven dimensions instead of the familiar four dimensions we know, allows for the possibility of parallel universes, as do Tegmark's probability calculations. Whether we could ever reach, observe, or even have definite knowledge of parallel worlds is an entirely different story.

Humankind Wiped out by Plague

SCIENCE FICTION TENDS to envision two possible scenarios in which catastrophic plague wipes out a large portion of Earth's population.

The first is deliberate, taking place as a result of biochemical weapons or other biowarfare. In Lester del Rey's "The Faithful" (1938), biological warfare exterminates all human life. Biological weapons also threaten to decimate humanity in Robin Cook's *Outbreak* (1987).

Biochemical warfare already exists: mustard gas in World War I and anthrax scares in the U.S. following 9/11. With the second war in Iraq and continued threats of global terrorism, the danger of widespread biowarfare has increased, and so has U.S. spending in this area: under the Pentagon's proposed 2005 budget, money for defense procurement would rise from $75 billion in 2005 to $114 billion by 2009, an increase of fifty-two percent.

In the second scenario, the plague is natural, or at least, not malicious—for instance, a new strain of flu, or germs that become resistant to antibiotics.

In Stephen King's novel *The Stand* (1978), most of the Earth's population is wiped out by a superflu known as "Captain Tripps." Survivors group into small communities and attempt to keep civilization going. King first explored this theme in his 1974 short story "Night Surf," of which *The Stand* seems to be an expanded version. The narrator of the story neatly sums up the situation:

> So here we were, with the whole human race wiped out, not by atomic weapons, or biowarfare or pollution or anything grand like that. Just the flu. I'd like to put down a huge plaque somewhere, in the Bonneville Salt Flats, maybe, Bronze Square. Three miles on a side. And in big raised letters it would say, for the benefit of any landing aliens: JUST THE FLU.

In the 1971 movie *The Omega Man*, based on Richard Matheson's 1954 short story "I Am Legend," a virus escaping from a military laboratory turns most of the people on Earth into living vampires. The population mutates, and the mutated plague victims drink human blood and sleep by day, being unable to stand bright light of any kind.

A plague virus taken from a laboratory also turns a large portion of the population into crazed, bloodthirsty killers in the movie *28 Days Later* (2002). The transformation takes place within seconds after the virus enters your bloodstream, most often as a result of being bitten by one of the victims.

Could a devastating new disease or plague wipe out a large segment of the human population? It has already happened on several occasions. And there is every indication it could happen again. Just look at the list of diseases plaguing us today: AIDS. Ebola. Swine flu. Toxic shock syndrome. Legionnaire's disease. And this is after we finally all but eliminated tuberculosis, influenza, malaria, scurvy, rickets, and smallpox. Of these, AIDS is the biggest worry—for now. No one has yet developed an effective vaccine or a cure. More than twenty-five million people have already died from AIDS, and over forty million more are infected today with the HIV virus.

New viruses are emerging with alarming frequency. In 1994, the first outbreak of a new type of virus, the Hendra, caused respiratory and kidney failure resulting in the death of two Australians. Another new virus, the Nipah virus, killed 105 people in Malaysia and Singapore during its first outbreak in 1998. Another outbreak of Nepah virus took place in the winter of 2003–2004, killing seventy-four percent of victims who became infected.

Even the plagues that have been tamed can potentially be unleashed again. The smallpox virus, for instance, was declared eradicated in the late 1970s. Yet a few remaining vials of the virus survive in the U.S. and Russia, stored in laboratories under secure conditions. During the Cold War, the U.S. and Soviet Union cultivated smallpox viruses and other deadly diseases as potential bioweapons. If they are somehow stolen and unleashed by terrorists, researchers at Johns Hopkins believe that smallpox would kill a million people within ninety days. Worse news: the World Health Organization is going to allow scientists to genetically modify some of these remaining samples of the smallpox virus, making it entirely possible that a new strain developed from this tinkering may be immune to the smallpox vaccine.

"We have decided to keep the beast alive in its cage," says an editorial in *NewScientist* (November 20, 2004). "Now we want to poke it." D. A. Henderson of the University of Pittsburgh, who led the drive to wipe out smallpox, called the planned experiments pointless and dangerous.

More bad news: diseases that we used to control with antibiotics are evolving to become resistant to the drugs. According to the *New England Journal of Medicine,* researchers have found bacteria in patients that can resist all currently available antibiotic drugs—making them virtually unstoppable.

In 1953, a virulent strain of the Staphylococcus auereus bacteria called 80/81 caused serious outbreaks of pneumonia worldwide. The bacteria was resistant to penicillin, but was virtually eliminated in the 1960s with methicillin and other new antibiotics. However, a new strain of the bacteria, which has recently been found in hospitals, is resistant to methicillin—and difficult to treat even with other antibiotics.

History is filled with horror stories of plague. In the fourteenth century, the Black Death—or Black Plague—killed two-thirds of China's population, one-third of Egypt's, and one-third of Europe's, including more than half of London's. Philip Zeigler describes the scene in his book *The Black Death* (1995):

> Many died daily or nightly in the public streets; of many others, who died at home, the departure was hardly observed by their neighbors, until the stench of their putrefying bodies carried the tidings; and what with their corpses and the corpses of others who died on every hand the whole place was a sepulcher.
>
> It was the common practice of most of the neighbors, moved no less by fear of contamination by the putrefying bodies than by charity towards the deceased, to drag corpses out of the houses with their own hands, aided, perhaps, by a porter, if a porter was to be had, and to lay them out in front of the doors, where any one that made the round might have seen, especially in the morning, more of them than he could count; afterwards they would have biers brought up or in default, planks whereon they laid them.
>
> Nor was it only once or twice that one and the same bier carried two or three corpses at once; but quite a considerable number of such cases occurred, one bier sufficing for husband and wife, two or three brothers, father and son, and so forth.
>
> And times without number it happened that, as two priests bearing the cross were on their way to perform the last office for someone, three or four biers were brought up by the porters in rear of them,

so that, whereas the priests supposed that they had but one corpse to bury, they discovered that there were six, or eight, or sometimes more.

One especially terrifying modern plague is necrotizing fasciitis, commonly known as the flesh-eating bacteria. Jacqueline Roemmele and Donna Batdorff describe the symptoms in their book *Surviving the "Flesh-Eating Bacteria"* (2000):

> As symptoms progress, the affected area continues to swell and becomes purple or mottled (blotches of black, purple, and red) in appearance, and a rash of blisters may appear. The rash then begins to spread to adjoining areas of the body. The skin may take on the texture of an orange peel.
>
> About one to three days after the advanced symptoms begin, the affected area swells to several times the normal size and may split open, discharging large amounts of thin, cloudy drainage fluid resembling dishwater.
>
> Large blisters (called bullae) filled with a bloody or yellowish fluid appear on the affected limb. Blackened necrotic lesions appear, causing the skin to break open. The patient's vital internal organs, such as kidneys, liver, and lungs, shut down due to toxic shock. The skin and other tissues can become blackened and slough off the body. Death follows shortly.

In her book *The Coming Plague* (1994), Laurie Garrett posits that a plague equal in intensity to Captain Tripps may be inevitable unless the scientific community can act quickly:

> As the Homo sapiens population swells, surging past the six billion mark at the millennium, the opportunities for pathogenic microbes multiply. If, as some have predicted, one hundred million of those people might then be infected with HIV, the microbes will have an enormous pool of walking immune-deficient Petri dishes in which to thrive, swap genes, and undergo endless evolutionary experiments.
>
> But as the world approaches the millennium, it seems, from the microbes' point of view, as if the entire planet, occupied by nearly six billion mostly impoverished Homo sapiens, is like the city of Rome in five B.C.
>
> "The world really is just one village. Our tolerance of disease in

any place in the world is at our peril," Joshua Lederberg said. "Are we better off today than we were a century ago? In most respects, we're worse off. We have been neglectful of the microbes, and that is a recurring theme that is coming back to haunt us."

While the human race battles itself, fighting over ever more crowded turf and scarcer resources, the advantage moves to the microbes' court. They are our predators and they will be victorious if we Homo sapiens do not learn how to live in a rational global village that affords the microbes few opportunities. It's either that or we brace ourselves for the coming plague.

Prehistoric Creatures, Alive in Modern Times

I N HIS 1912 NOVEL *The Lost World*, Sir Arthur Conan Doyle, creator of Sherlock Holmes, tells the tale of a jungle expedition which seeks—and finds—prehistoric creatures thought to have become extinct ages ago. The explorers eventually become trapped in the jungle and are stalked by hungry, carnivorous dinosaurs. In their first encounter with the beasts inhabiting the Lost World, a pterodactyl steals their dinner:

> Suddenly out of the darkness, out of the night, there swooped something with a swish like an aeroplane. The whole group of us were covered for an instant by a canopy of leathery wings, and I had a momentary vision of a long, snake-like neck, a fierce, red, greedy eye, and a great snapping beak, filled, to my amazement, with little, gleaming teeth.
>
> The next instant it was gone—and so was our dinner. A huge black shadow, twenty feet across, skimmed up into the air; for an instant the monster wings blotted out the stars, and then it vanished over the brow of the cliff above us.

Edgar Rice Burroughs, creator of Tarzan, writes about a similar theme in his 1924 novel *The Land That Time Forgot*. In that novel, he tells the tale of explorers who accidentally stumble into a region that seems not to have evolved, and is the same as it was in prehistoric times:

> We were in the middle of a broad and now sluggish river. Close by us something rose to the surface of the river and dashed at the periscope. Above the trees there soared into my vision a huge thing on batlike wings—a creature large as a large whale, but fashioned more after the order of a lizard.
>
> All about us was a flora and fauna as strange and wonderful to us as might have been those upon a distant planet had we suddenly been miraculously transported through ether to an unknown world. Even

the grass upon the nearer bank was unearthly—lush and high it grew, and each blade bore upon its tip a brilliant flower—violet or yellow or carmine or blue—making as gorgeous a sward as human imagination might conceive.

The tall, fernlike trees were alive with monkeys, snakes, and lizards. Huge insects hummed and buzzed hither and thither. Mighty forms could be seen moving upon the ground in the thick forest, while the bosom of the river wriggled with living things, and above flapped the wings of gigantic creatures such as we are taught have been extinct throughout countless ages.

As much as these discoveries of prehistoric lands in present time tickle our imagination, it has never happened. No one has ever found a region where evolution has stood still and the flora and fauna is the same as it was thousands of years ago.

We have, however, discovered at least one prehistoric creature alive and well today, millions of years after it first appeared on Earth: a large fish called the coelacanth. Samantha Weinberg describes this amazing discovery in her book, *A Fish Caught in Time* (2000):

> How, I wondered, had it survived, virtually unchanged, for all that time, lost at the bottom of the vast, cold ocean, watching silently as other creatures evolved and became extinct? Homo sapiens first walked on earth only a hundred thousand years ago; the fish before me, suspended in murky formalin, pre-dated modern man by 399.9 million years.
>
> On November 12, 1954, at 8 P.M., Zema ben Said caught the coelacanth, 1,000 meters off Mutsamudu, on the northwestern coast of Anjouan. By 9:30 P.M., the makeshift aquarium was ready.
>
> The live coelacanth was released into it, and the boat was covered by a net to prevent the fish from escaping. Every half an hour, the boat was rocked to enable fresh water to enter. Witnesses reported a dark grayish-blue fish, the color of watch-spring steel, with luminescent green-yellow eyes.
>
> That night, the population of Mutsamudu celebrated the valuable catch. They sang and danced until daybreak, making regular trips to view their precious fish. The coelacanth at first appeared bewildered but otherwise calm; it swam slowly, by curious rotating movements of its pectoral fins, using its second dorsal fin, anal fin, and tail as a rudder.

Alligators and crocodiles, as well as sharks, also existed, albeit in slightly different form, in prehistoric times, and they, too, are still with us today. For instance, giant crocodiles reaching lengths of forty feet once roamed the Earth, and dinosaurs were frequently their prey. Today's crocodiles are nearly identical to these beasts, except most are less than one-third their size.

Some investigators speculate that the Loch Ness monster, like the coelacanth, is a species of sea creature from prehistoric times that somehow became trapped in the Loch Ness Lake into which it came from the ocean. Could a more advanced species from prehistoric times survive into the modern era? In his 1952 short story, "Sail On! Sail On!" Philip José Farmer wonders what it would be like if a few remaining Neanderthals had survived and made it into the twentieth century.

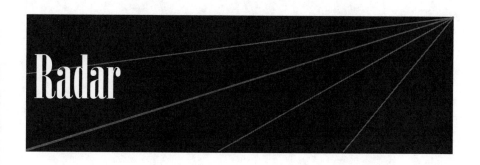

Radar

H UGO GERNSBACK predicted radar, complete with diagrams, in *Ralph 124C 41+*—originally published as a series of chapters in Gernsback's magazine *Modern Electrics* (1911–1912) and then as a novel in 1925. The term radar is used in the title of the film *Radar Men from the Moon* (1952), in which astronauts journey to the Moon to battle aliens plotting to conquer the Earth.

Radar, an acronym for "radio detection and ranging," uses radio waves to detect airborne objects and determine their distance. In 1935, Scottish physicist Robert Alexander Watson-Watt built a radar that could follow the path of an airplane by the microwave reflections it sent back (microwaves are nothing more than a radio waves with short wavelengths).

Calculating the distance of the airplane is easy. You measure the time it takes for your signal to be transmitted, reflected off the plane, and bounced back to you. Divide that time by two (you are interested in the one-way distance of the plane to your radar station, not the total distance of the round-trip route), then multiply by the speed of the microwaves. Underwater, sonar devices use ultrasonic sound instead of radio waves to similarly determine the distance, location, and even shapes of submerged objects in relation to a boat or submarine.

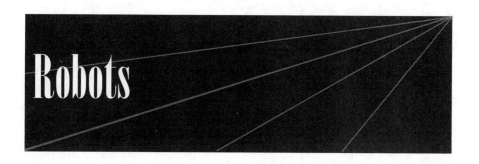

Robots

KAREL CAPEK'S 1920 PLAY *R.U.R.* (*Rossum's Universal Robots*) was the first science fiction story to feature robots, although Capek's robots were flesh and blood creatures. Another robot, the Tin Man, is a key character of Frank Baum's classic fantasy story, *The Wizard of Oz* (1900).

But the notion of robots has its roots in Greek mythology. Hephaestus, the Smith-god, built golden mechanical women to help him in his workshop; they could operate the bellows and even talk. Hephaestus also built artificial men to guard the island of Minos.

In the nineteenth century, E. T. A. Hoffman wrote a couple of stories about automata (an "automaton" is a being that behaves in a mechanical fashion, like the mechanical figures used in many Disneyworld rides and displays): "Automata" (1814) and "The Sandman" (1816). Herman Melville wrote about a murderous automaton in "The Bell Tower" (1855).

Lester del Rey's "Helen O'Loy" (1938) offers a robot as a perfect wife. Robert Moore Williams' "Robots Return" (1938), Joseph E. Kelleam's "Rust" (1939), and Eando Binder's "The Robot Aliens" (1935) were early pulp-magazine stories.

Robots appeared frequently in 1950s science fiction films. In *Colossus of New York* (1954) a man's brain is placed into the body of a twelve-foot-tall robot, creating, in effect, a cyborg. In *Tobor the Great* (1954) a robot is built to withstand the stress of space travel.

Several cartoons in the 1960s, including *Tobor the 8ᵗʰ Man* and *Gigantor*, also featured robots. Gigantor was a giant robot controlled by a young boy. Tobor was an android who smoked "atomic cigarettes" to recharge his atomic power source and had a spare brain in his shoulder in case his head became damaged.

The most famous science fiction robot from the movies is perhaps Robby the Robot from the 1956 film *Forbidden Planet*. Robby made repeat appearances in subsequent sf films and "guest-starred" in the

1965–1968 television series *Lost in Space*. The *Lost in Space* robot (a model B-9) was best known for his alert, delivered in a monotone: "Warning! Warning!"

Inventors, engineers, and scientists have been designing and building mechanical beings for decades. In 1769, centuries before IBM's Big Blue computer beat the human world champion at chess, Wolfgang von Kempelen, a Hungarian inventor, built a chess-playing robot, although he cheated: a person was actually inside the machine making the moves. Here's how Gaby Wood describes the device in her book *Edison's Eve* (2002):

> It was a machine that seemed to think ... the Chess Player was a carved wood figure of a man in Turkish garb (white turban, striped silk shirt, fur-trimmed jacket, white gloved hands, drooping mustache) sitting at a wooden chest, to which was affixed a large chessboard. The measurements given for the chest were between 4 and 4 ½ feet long, about 2 feet deep and between 3 and 3 ½ feet high.
>
> The player's right hand rested flat on the top, next to the chessboard. In his left hand, slightly raised from the surface, he held a long thin pipe. His head moved on his neck, his eyes moved in their sockets. He had the appearance, as one visitor put it, "of someone who has just been smoking."

In many science fiction stories, three laws governing the behavior of all robots are programmed into every robot's brain during its manufacture. The Three Laws of Robotics, invented by Isaac Asimov, are as follows:

1. A robot may not injure a human being, or through inaction, allow a human being to come to harm.
2. *A robot must obey the orders given it by human beings, except where such orders would conflict with the First Law.*
3. A robot must protect its own existence, as long as such protection does not conflict with the First or Second Law.

Eando Binder had written a story called "I, Robot" in 1939. Isaac Asimov appropriated Binder's title for his first collection of robot stories, *I, Robot* (1950), the first of which was "Strange Playfellow" (1940, aka "Robbie"). Asimov later followed with numerous other robot stories and novels, including the Lije Bailey and R. Daneel Olivaw novels beginning with *The Caves of Steel* (1954). Two of Asimov's stories have been turned into films in the past few years—*The Bicentennial Man* (1999) and *I, Robot* (2004). Asimov claims that Joseph Engelberger, who in the late 1950s founded Unimation, a firm devoted to the design and production of robots, says he became interested in robotics after reading *I, Robot.* Other robot-based sf stories include:

- Harry Bates' "Farewell to the Master" (1940).
- Eric Frank Russell's "Jay Score" (1941).
- Anthony Boucher's "Q.U.R." (1942).
- Jack Williamson's "With Folded Hands" (1947).
- Clifford Simak's *City* (1952).
- Henry Kuttner's *Robots Have No Tails* (1952).
- Walter Miller, Jr.'s "The Darfsteller" (1955).
- Stanislaw Lem's *The Cyberiad* (1974).
- Roger Zelazny's "For a Breath I Tarry" (1966).
- Robert Sheckley's "The Cruel Equations" (1971).
- Robert Silverberg's "Good News from the Vatican" (1971).
- Walter Tevis' *Mockingbird* (1980).
- John T. Sladek's *Roderick* (1980).
- Tanith Lee's *The Silver Metal Lover* (1981).

Walt Disney was enamored with robotic technology, and Disney World is full of animatronic figures—robots that look and sound like real people and animals. An Abe Lincoln automaton in the Hall of Presidents recites the Gettysburg address. The Tiki Room is populated by 225 animatronic animals and birds; the "It's a Small World" and "Pirates of the Caribbean" rides also feature dozens of these figures.

The first industrial robot was patented in 1954 by Joe Engelberger, who formed a company, Unimation, to sell and market them. The device was a programmable robot arm controlled by a computer. In 1961, General Motors became the first manufacturer to install a robot in one of its factories.

Today, of course, robots are commonplace in factories, where specialized robots perform many tasks on assembly lines. More than eight hundred thousand robots perform valuable tasks at industrial and manufacturing facilities worldwide. Robots are ideal for assembly line or fabrication work, where each step in the process is a simple task that must be repeated accurately, swiftly, and constantly throughout the work shift. For example, the Flex Picker, a robot made by ABB, picks up objects off a conveyor belt and places them into a carton.

Manufacturers are also introducing home versions into the market like Honda's Asimo, which can walk forward and backward, turn corners, and go up and down stairs. In development are Asimo models capable of understanding voice commands, carrying loads, and doing simple household chores. A report from the United Nations predicts that the number of robots people have in their homes will increase by four million worldwide from 2004 to 2007.

There is an ongoing debate about whether robots, automation, and computers are good or bad for the human race. Advocates of automation argue that a machine like the Flex Picker handles boring, mind-numbing repetitive tasks, freeing humans to concentrate on more creative work.

Opponents say that such machines are in fact putting a lot of people permanently out of work, many of whom are too old or inflexible to be trained for other jobs. And there are probably some people of modest or lower intelligence for which a repetitive factory job is just the right employment for them; not everyone wants to—or can—program computers or write novels.

Building robots has become a popular hobby for amateur scientists and engineers. On the TV show *Battlebots*, contestants avoid the ring themselves and send their robots in to fight; the objective is to disable or destroy the other person's robot. The robots, which can weigh hundreds of pounds (weight is an advantage; you can run your own robot into the other and try to damage him through collision), are equipped with such weapons as rotary saws, spikes, hammers, and forklift-style prongs for flipping the other robot on its back.

And what of the future? In his book *Robot* (1999), Hans Moravec outlines the four stages of development he predicts will be the evolution of the robot over the next four decades:

1. **First-Generation Universal Robots.** Estimated time of arrival: 2010. Processing power: 3,000 million instructions per second (MIPS). Distinguishing feature: General-purpose perception, manipulation, and mobility.

2. **Second-Generation Universal Robots.** Estimated time of arrival: 2020. Processing power: 100,000 MIPS. Second-generation robots, with thirty times the processing power of the first generation, will learn on the job. Their big advantage is adaptive learning, which "closes the loop" on behavior. Each robot action is repeatedly adjusted in response to measurements of the action's past effectiveness.

3. **Third-Generation Universal Robots.** Estimated time of arrival: 2030. Processing power: 3 million MIPS. A third generation of universal robots will have onboard computers as powerful as the supercomputers that optimized second-generation robots. They will learn much faster than the first and second generations because they do much of their trial and error in fast simulation rather than rather than learn through physical action, which is far slower.

4. **Fourth-Generation Universal Robots.** Estimated time of arrival: 2040. Processing Power: one hundred million MIPS. Fourth-generation universal robots will have computers powerful enough to simultaneously simulate the world and reason about the simulation. They will also be able to understand human languages.

Scientists in Colorado are working to develop a robotic brain with 3.7 million artificial neurons, each consisting of a group of transistors contained in a cell. The researchers are hoping that the artificial brain will allow a robot to interact with stimuli in its environment and develop the sort of intelligence seen in animals.

Sony has created Aibo, an artificial dog with a CPU for a brain and sensors providing the feedback of eyes and ears. Pressure sensors enable Aibo to detect people petting him. Eyes with a built-in 180,000-pixel miniature color video camera can recognize objects. Software allows Aibo to perform complex actions, display the semblance of emotions and instincts, and learn through experience.

Robotic Surgery

ROBOTIC SURGERY IS PRECISELY what its name indicates: operations are performed either by robot surgeons or human surgeons aided by robotic tools.

Neil R. Jones' Professor Jameson stories, the first of which is "The Jameson Satellite" (1931), feature a robot civilization that discovers the frozen body of Professor Jameson in a spaceship; the robots surgically transplant his revived brain into a robot body.

In Jack Williamson's "With Folded Hands" (1947), robots called "humanoids" perform a lobotomy on Mr. Sledge. And in Isaac Asimov's "The Bicentennial Man" (1976), surgery is routinely performed by robot surgeons whose hands are far steadier and more precise than human physicians.

Today, "robotic surgery" has indeed become a reality. Robotic surgery is now being performed at hundreds of hospitals in the U.S. and around the world. But unlike the surgical robots in Isaac Asimov's stories, today's surgical robots still need human operators. Instead of replacing human doctors, they enable the surgeons who use these robotic devices to be faster and more precise, and even to operate on patients by remote control.

When using a robotic surgical system, a doctor works several feet away from the patient and remotely manipulates robotic arms to do the work. The doctor sits at a console, looking at three-dimensional images of the patient on a magnified viewfinder. The human doctors perform the surgery by looping their fingers around controls on the console—which translates their hand movements to the robot's mechanical "hands." Because doctors can operate while sitting at a console, robotic surgery is less tiring and error-prone than conventional surgery. With only a keyhole incision, the robot-aided surgeon can perform intricate operations that are traditionally done by opening up the patient. Other advantages of robot-aided surgery include smaller incisions, less bleeding, faster patient recovery, fewer scars, reduced patient risk, shorter duration of surgery, and less infection.

An aging U.S. population is the real reason why robotic surgery will transition from a novelty today to "standard operating procedure" within the next three to five years.

As people age, the risk of side effects from surgery multiplies. Robotic surgery can reduce those risks to acceptable levels. For instance, there are 230,000 new cases of prostate cancer discovered every year in the U.S., but only 80,000 people undergo surgery. That's because open surgery carries a high degree of risk.

But now, five percent of the radical prostatectomies in the U.S. are done using robotics with a minimally invasive procedure that dramatically reduces the risk of impotence and incontinence—the two great fears of patients facing surgery.

In short, robotic surgery—once a science fiction concept—is rapidly becoming "standard operating procedure" in modern medicine. And now, even robotic nurses are getting into the picture. At New York-Presbyterian Hospital, doctors use a device called the Penelope Surgical Instrument Server as a "robotic surgical nurse." The robot hands give doctors the surgical instruments they request on command, and then return the instruments to their original place once the surgeon has finished with them.

Rocket Ships

ALTHOUGH THERE HAVE BEEN science fiction stories based on physics, chemistry, mathematics, biology, and even sociology, astronomy is at the top of the list of science disciplines that inspire science fiction. And of course, to get to outer space, you need a rocket ship.

The Chinese built the first rockets nine centuries ago, but these were basically fireworks, not designed for transportation or to carry a crew. In the thirteenth century, the Chinese used rockets as weapons of war. During the War of 1812, Colonel William Congreve, an English officer, used larger versions of such rockets to drive away American troops at the battle of Bladensburg.

One of the first scientists to take rockets seriously was Konstantin Tsiolkovsky, a Russian physicist. As early as 1895 he began writing papers in which he discussed liquid fuel rockets, space flights, space suits, satellites, space stations, and the colonization of other planets. He also presented these ideas in his science fiction novel *Outside the Earth* (1920).

One of the earliest stories to feature a rocket ship was Cyrano de Bergerac's *A Voyage to the Moon* (1662). He describes a primitive rocket ship as a cedar coffer equipped with a "burning glass" at one end. The glass heats the air inside the coffer by focusing the Sun's rays, and the gas is expelled through a hole at the other end to provide propulsion. In Edgar Allen Poe's "The Unparalleled Adventures of One Hans Pfaal" (1835), the hero starts his balloon journey to the Moon by a blast of rockets.

Rockets were prominently featured in Ray Bradbury's Martian Chronicles sequence of stories, most notably his 1947 story "Rocket Summer." That same year, Robert Heinlein published his novel *Rocket Ship Galileo*.

No one writes more poetically about rockets and space travel than Bradbury. In "Kaleidoscope," an astronaut is thrown into outer space when his ship explodes. His spacesuit keeps him alive until, pulled in by

Earth's gravity, he is incinerated as he enters the atmosphere; a small boy seeing this thinks the astronaut is a shooting star and makes a wish.

In Bradbury's "The Rocket Man," (1951) a family loses the father, a spaceman, when his ship falls into the Sun. And in "No Particular Night or Morning" (1951), the loneliness of outer space drives an astronaut insane.

Numerous other science fiction novels and stories involving rockets and space travel have been published. These include:

- Jack Williamson's *The Legion of Space* (1934).
- Edmond Hamilton's "What's It Like Out There?" (1952).
- Fredric Brown's *The Lights in the Sky Are Stars* (1953).
- Frederik Pohl's *Gateway* (1977).
- Gregory Benford's *In the Ocean of Night* (1977).
- Ben Bova's *Voyagers* (1981).
- Charles Pellegrino's *Flying to Valhalla* (1993).
- Jack McDevitt's *Ancient Shores* (1996).

Robert Goddard launched the first successful liquid fuel rocket in 1926. He was reputedly inspired by H. G. Wells' *The War of the Worlds*, published in 1898. John Munro's *A Trip to Venus* (1897) included a method of liquid-fuel staged rockets.

Stanley Beitler describes how liquid fuel rockets work in his book *Rockets and Your Future* (1961):

The liquid rocket [has] two separate tanks—one for fuel, the other for the oxidizer to burn the fuel. The liquid rocket's highly volatile propellant must be pumped in just before takeoff, a time-consuming process when seconds are precious. However, this elaborate design of pipes, valves, etc., has its advantages.

For one, it permits the engine to be cooled by the propellant it carries, and secondly, the airframe can be made of this metal, resulting in a significant weight reduction. Also, this plumbing allows the rate of flow of the propellant to be easily adjusted to furnish various amounts of thrust.

In the liquid-fueled rocket, the fuel and the oxidizer (usually liquid oxygen) are forced into the combustion chamber under high pressure and upon contact explode and burn; the resultant gas particles are discharged through the nozzle.

Temperatures created in the combustion chamber reach as high as

4000° to 6000° F., which can melt the structure of the chamber and nozzle. This is prevented by circulating either the liquid oxidizer or the fuel through the hollow walls of the combustion chamber and nozzle prior to introducing them into the combustion chamber for burning. This cycle is continuous and is called regenerative cooling.

The problem with today's liquid and solid rockets is that they must carry an enormous weight of chemical fuel to accelerate to a speed where they can escape Earth's gravity. To solve the problem, sf writers propose other propulsion methods, such as the ion drive described by Poul Anderson in his 1970 novel *Tau Zero*:

The ion drive came to life. Reaction mass entered the fire chamber. Thermonuclear generators energized the furious electric area that stripped those atoms down to ions; the magnetic fields that separated positive and negative particles; the forces that focused them into beams; the pulses that lashed them to even higher velocities as they sped down the rings of the thrust tubes; until they emerged scarcely less fast than light itself.

Their blast was invisible. No energy was wasted on flames. Instead, everything that the laws of physics permitted was spent on driving Leonora Christine outward.

Another alternative to solid or liquid fuel chemical rockets was described by Arthur C. Clarke in his 1963 short story "The Wind from the Sun." Spaceships in the story are equipped with giant sails that enable the craft to be propelled by the "solar wind"—the stream of particles continually emanating from the Sun:

Hold your hands out to the Sun. What do you feel? Heat, of course. But there's pressure as well—though you've never noticed it, because it's so tiny. Over the area of your hands, it comes to only about a millionth of an ounce.

But out in space, even a pressure as small as that can be important, for it's acting all the time, hour after hour, day after day. Unlike rocket fuel, it's free, and unlimited. If we want to, we can use it. We can build sails to catch the radiation blowing from the Sun.

You can see how light it is. A square mile weighs only a ton, and can collect five pounds of radiation pressure. So it will start moving—and we can let it tow us along, if we attach rigging to it.

Of course, its acceleration will be tiny—about a thousandth of a g. That doesn't seem much, but let's see what it means.

Please remember that in space there's no friction; so once you start anything moving, it will keep going forever. You'll be surprised when I tell you what our thousandth-of-a-g sailboat will be doing at the end of a day's run: almost two thousand miles an hour! And all without burning a single drop of fuel!

In several essays on rocketry and space flight, Clarke suggests that spaceships could be launched from a Moon base instead of Earth. Since the Moon's gravity is one-sixth of Earth's, far less fuel and thrust is required to escape it. Robert Heinlein describes a lunar rocket launch in his 1952 novel *The Rolling Stones*:

Blasting off from Luna is not the terrifying and oppressive experience that a lift from Earth is. The Moon's field is so weak, her gravity well so shallow, that a boost of one-g would suffice—just enough to produce Earth-normal weight.

Captain Stone chose to use two gravities, both to save time and to save fuel by getting quickly away from Luna—"quickly" because any reactive mass spent simply to hold a spaceship up against the pull of a planet is an "overhead" cost; it does nothing toward getting one where one wants to go.

Furthermore, while the *Rolling Stone* would operate at low thrust she could do so only by being very wasteful of reactive mass, i.e., by not letting the atomic pile heat the hydrogen hot enough to produce a really efficient jet speed.

So he caused the *Stone* to boost at two gravities for slightly over two minutes. Two gravities—a mere nothing! The pressure felt by a wrestler pinned to the mat by the body of his opponent; the acceleration experience by a child in a schoolyard swing—hardly more than the push resulting from standing up very suddenly.

Smart Weapons

AMES BLISH'S "TOMB TAPPER" (1956) imagines guided missiles, and A. E. van Vogt's *The Weapon Shops of Isher* (1951) deals with powerful hand weapons.

For our purposes, let's define a "smart weapon" as a gun that fires only in the hands of the owner. The judges in the 1995 movie *Judge Dredd*, starring Sylvester Stallone, carry powerful multifunction hand weapons, know as "law givers." The law giver responds to voice commands, but will only fire when it recognizes the fingerprints of the judge authorized to use it. The problem with smart weapons is that there is always a way to fool them: the villain in the movie, played by Armand Assante, steals Dredd's law giver and is able to fire it. The reason: he is a clone of Dredd, with identical fingerprints.

At the New Jersey Institute of Technology, researchers have developed a sensor-based hand grip similar to the law giver's. But instead of detecting the user's fingerprint, it recognizes his unique grip as his fingers tighten around the trigger. The sensor detects the size of the hand, length of each individual segment in the fingers, and pressure of the grip at all points. A microchip compares this data to the user's measurements. If they do not match, the weapon does not fire.

There are other ways to match the owner with the weapon and ensure that a stranger cannot fire it. In one smart weapon design, the owner wears a magnetic ring that matches a magnetic inside the gun, enabling it to be fired. Similarly, the owner can carry a transponder that activates the gun when it is within close range.

Space Elevator

I N HIS 1979 NOVEL *The Fountains of Paradise*, Arthur C. Clarke suggests building a tower of sorts—actually an elevator—that reaches literally from the Earth's surface to 23,500 miles above it, with the last stop at the top being outer space itself. He also wrote about the possibilities of space elevators in a 1981 technical paper. In Clarke's invention, a cable system is attached to the Earth at the equator. The cable extends upward to the point where its center of gravity maintains a geostationary orbit. That center of gravity provides an upper platform for electric cars running up and down along the cable. The elevator could carry satellites and other equipment beyond Earth's gravity, eliminating the need for rocket launches.

Now it appears that NASA is ready to help make Clarke's dream of a quick and easy pathway to outer space a physical reality: in 2005, NASA announced that it would award a $100,000 prize to the firm that comes up with the best design for an actual space elevator. The space elevator will carry astronauts, equipment, and supplies from Earth's surface to a space station in geosynchronous orbit at an altitude of approximately twenty-three thousand miles.

The elevator cable will be made of carbon nanotubes, chosen because they are up to ten times stronger than steel but much lighter in weight. In fact, it was not until the discovery of carbon nanotubes that engineers began to think Clarke's space elevator could become a reality. Here's how Clarke describes it in his novel:

> ...it's a continuous pseudo-one-dimensional diamond crystal—though it's not actually pure carbon. There are several trace elements in carefully controlled amounts. It can be mass-produced only in the orbiting factories, where there's no gravity to interfere with the growth process....
>
> Our design is a hollow square tower with a track up each face. Think of it as four vertical railroads.... Where it starts from orbit, it's

forty meters on a side, and it tapers down to twenty when it reaches earth....

A good analogy now would be the old Eiffel Tower—turned upside down and stretched out a hundred thousand times. The reason you can't see this sample is that it's only a few microns thick. Much thinner than a spider's web.

Electricity for the elevator's motor will be beamed up by laser or microwaves, since a copper cable would be too heavy. Building the space elevator will probably cost $10 billion and take fifteen years.

Subsurface Life

W E PRIDE OURSELVES on having explored our planet, but at best we've only scratched the surface, quite literally, for the vast interior of Earth is still almost wholly unknown to us," writes Terry Carr in *Creatures from Beyond* (1975). "What creatures might dwell there, burrowing inexorably through ancient rock? And if one of these creatures should come up to the surface, what then?"

In the 1864 Jules Verne novel *Journey to the Center of the Earth*, explorers find a pathway leading to a center of the Earth, and when they get there, they discover a whole other world populated by humans, flora, and fauna, complete with its own ocean and giant sea monsters:

"It is a colossal monster!" I cried, clasping my hands.

"Yes," cried the agitated Professor, "and there yonder is a huge sea lizard of terrible size and shape."

"And farther on behold a prodigious crocodile. Look at his hideous jaws, and that row of monstrous teeth. Ha! He has gone."

"A whale! A whale!" shouted the Professor, "I can see her enormous fins. See, see, how she blows air and water!"

We stood still—surprised, stupefied, terror-stricken at the sight of this group of fearful marine monsters, more hideous in the reality than in my dream.

They were of supernatural dimensions; the very smallest of the whole party could with ease have crushed our raft and ourselves with a single bite.

The fearful reptiles advanced upon us; they turned and twisted about the raft with awful rapidity.

On one side the mighty crocodile, on the other the great sea serpent. The rest of the fearful crowd of marine prodigies have plunged beneath the briny waves and disappeared!

Edgar Rice Burroughs, the prolific pulp writer, sent Tarzan on a journal to the center of the Earth in his 1930 novel *Tarzan at the Earth's Core*. Subterranean Earth in the Burroughs books is richly populated with flora, fauna, and an entire civilization of humans. While Verne's intrepid explorers journeyed to the Earth's core on foot, Tarzan was lazier and rode into the subterranean world in a ship built by an inventor friend of his, Jason Gridley:

Eight air-cooled motors drove as many propellers, which were arranged in pairs upon either side of the ship and staggered in such a manner that the air from the forward propellers would not interfere with those behind.

The engines, developing 5,600 horsepower, were capable of driving the ship at a speed of 105 miles per hour.

The ordinary axial wire, which passes the whole length of the ship through the center, consisted of a tubular shaft of Harbenite from which smaller tubular braces radiated, like the spokes of a wheel, to the tubular girders, to which the Harbenite plates of the outer envelope were welded.

Owing to the extreme lightness of Harbenite, the total weight of the ship was 75 tons, while the total weight of its vacuum tanks was 225 tons.

For purposes of maneuvering the ship and to facilitate landing, each of the vacuum tanks was equipped with a bank of eight air valves operated from the control cabin at the forward end of the keel; while six pumps, three in the starboard and three in the port engine corridors, were designed to expel the air from the tanks when it became necessary to renew the vacuum.

Special rudders and elevators were also operated from the forward control cabin as well as from an auxiliary position aft in the port engine corridor, in the event that the control cabin steering gear should break down.

The 2003 motion picture *The Core* featured a similar rocket-shaped ship for traveling into the Earth's core. The mission: to stave off the destruction off Earth by restarting the rotation of Earth's molten center by detonating nuclear explosives within it. Here's the designer of the ship explaining the drilling technology:

Concentrated sound waves fire out from *Virgil's* resonance chamber, crushing the rock in front of the ship. The laser melts the crushed rock, and that slag is pulled by the impeller blades through this tube and shoots out the back.

The ship's outer hull is fabricated from high-compression living molecules linked in a carbon matrix. As the ship drills deeper into the planet, the pressure aligns the molecules, increasing the hull's strength. The frame contracts under the pressure to reinforce the hull. The molecular "skin" of the vessel is not burned away by the intense heat of the Earth's core because the high temperatures cause new nanomicrobes to grow and replace the old ones faster than they can be incinerated.

Two episodes of the old *Superman* television show were shown in limited theater release as the 1951 movie *Superman and the Mole Men*. A crew drilling for oil deeper than an oil well has ever been drilled before breaks through the Earth's crust and hits empty space. Small, man-like creatures emerge through the hole and, although they mean no harm, terrorize the local residents.

Far more advanced than the primitive inhabitants of *Journey to the Center of the Earth*, the Mole Men had developed a technology that includes a pulsed energy weapon, which they use to defend themselves against the frightened locals. The device was actually an upright vacuum cleaner held horizontally.

In David H. Keller's 1929 story "The Worm," the vibrations from an old stone mill attract a monstrous giant worm who mistakes the vibration for the movement of a potential mate.

The giant worm, whose natural habitat is hundreds of feet below the surface, drills through solid rock by eating through it. From Keller's description of the creature, we can see that his worm is the precursor of the monster worms in the 1990 Kevin Bacon film *Tremors*, though even larger:

> There was a central mouth filling half the space; fully fifteen feet in diameter was that mouth, and the sides were ashen gray and quivering. There were no teeth.
>
> The circular lip seemed made of scales of steel, and they were washed clean with the water from the face.
>
> On either side of the gigantic mouth was an eye, lidless, browless, pitiless. They were slightly withdrawn into the head so that Thing could bore into rock without injuring them.

In the late 1950s, Project Mohole attempted to drill deeper than anyone had drilled before, through Earth's crust and into the mantle. The plan was to drill through the ocean floor, where the crust is thinnest. Five holes, one of which extended more than six hundred feet beneath the sea floor, were drilled at a depth of 11,700 feet under water. The project was discontinued in 1966 because of the cost, and the mantle was never reached.

Is there life deep beneath the planet's surface? What kind? A relatively new field of science, "deep biology"—the study of subsurface bacteria—is dedicated to exploring these questions.

In the late 1980s, scientists found microbes living in rock more than 1,600 feet below the surface in South Carolina. In 1996, Tullis Onstott, a geologist from Princeton University, found bacterial spores trapped in three-billion-year-old rock in South Africa's deepest gold mine, nearly two miles below the surface. In 1993, scientists from Texaco Oil Company found microbes living in rocks more than 1.5 miles underground.

No living things other than microbes have been found at these levels underground, for several reasons: lack of sunlight, lack of room to move and grow, lack of a food supply, and the extreme heat. As you go deeper in the Earth, the rock gets hotter because it is closer to the Earth's molten core. In the South African gold mine explored by Onstott, the temperatures reached a nearly unbearable 140 degrees Fahrenheit.

There is some evidence that living organisms may exist even deeper than a couple of miles underground. Thomas Gold, a physicist whose avocation is petroleum geology, reports finding magnetite, a reduced form of iron oxide, at the bottom of an oil well nearly four miles deep. Magnetite is often a sign of bacterial activity.

Or even deeper: tiny, single-celled organisms have been discovered living in the Marianas Trench, the deepest location in the Pacific Ocean, at a depth of seven miles. Despite the enormous, crushing pressures at this depth, the creatures, known as *foraminifera*, are soft-shelled. Scientists discovered a small colony of the organisms in a scoop of mud brought up by an unmanned submersible.

Finding ample energy presents something of a problem for subsurface life due to the lack of sunlight and food. Geobacter sulfurreducens, a family of bacteria that live underground, extend tiny filaments ten times the length of their body to touch iron oxide crystals, from which they strip electrons to provide themselves with energy.

Superbeings (see also "Bionics," "ESP," "Mutations")

EVER SINCE JERRY SIEGEL and Joe Shuster created Superman more than half a century ago, readers have been fascinated with the notion of beings with enhanced abilities beyond those of ordinary human beings.

In a *Twilight Zone* episode, Burgess Meredith plays Luther Dingle, a mild-mannered man who is endowed with superstrength by two aliens visiting Earth for the purposes of conducting experiments on human beings. At the end of the episode, the aliens remove Dingle's strength; but another pair of aliens endows him with another enhancement: superintelligence.

In the 1956 movie *Forbidden Planet*, humans discover a planet once occupied by a highly advanced race of aliens. Using one of the machines the aliens left behind, two of the men give themselves superintelligence—boosting their I.Q. beyond any level previously recorded.

In Daniel Keyes' classic 1959 story "Flowers for Algernon," a retarded man named Charlie Gordon has his I.Q. raised from sixty-nine to more than two hundred through an experimental surgical procedure; unfortunately, the increase in intelligence is not permanent:

> Strauss dealt largely with the theory and techniques of neurosurgery, describing in detail how pioneer studies on the mapping of hormone control centers enabled him to isolate and stimulate these centers while at the same time removing the hormone-inhibitor-producing portion of the cortex. He explained the enzyme-block theory and went on to describe my physical condition before and after surgery.
>
> The surgical stimulus to which we were both subject has resulted in an intensification and acceleration of all mental processes. The unforeseen development, which I have taken the liberty of calling the Algernon-Gordon Effect, is the logical extension of the entire intelligence speed-up. The hypothesis here proven may be described simply in the following terms: Artificially increased intelligence deteriorates at a rate of time directly proportional to the quantity of the increase.

Gully Foyle, the protagonist of Alfred Bester's 1956 novel *The Stars My Destination* attains, through surgery, the ability to accelerate his speed and reflexes by a factor of five:

He stripped and examined his body. He was in magnificent condition, but his skin still showed delicate silver seams in a network from neck to ankles. It looked as though someone had carved an outline of the nervous system into Foyle's flesh. The silver seams were the scars of an operation that had not yet faded.

That operation had cost Foyle a $200,000 bribe to the chief surgeon of the Mars Commando Brigade and had transformed him into an extraordinary fighting machine. Every nerve plexus had been re-wired, microscopic transistors and transformers had been buried in muscle and bone, a minute platinum outlet showed at the base of his spine. To this Foyle affixed a power-pack the size of a pea and switched it on. His body began an internal electronic vibration that was almost mechanical.

He backed a step and pressed his tongue against his upper incisors. Neural circuits buzzed and every sense and response in his body was accelerated by a factor of five.

The effect was an instantaneous reduction of the external world to extreme slow motion. Sound became a deep garble. Color shifted sown the spectrum to the red. The two assailants seemed to float toward him with dreamlike languor. To the rest of the world Foyle became a blur of action....

Special Agent Salinas, the hero of Allen L. Wold's 1980 novel *Star God*, is given special powers through operations and other procedures conducted by different races of aliens:

The skin of his legs, his stomach, his arms, as much of himself as he could see, was covered with a bright network of fine, shiny silver lines, so close together they seemed merely a texture. They ran in three directions, forming a triangular pattern all over his body. His normal skin color was lost in the silvery mesh, which made him glisten and twinkle like sunlight on shaking water.

"My skin, what's happened to it?'

"Nothing, except that it is better than it was. More resistant and resilient, stronger, more sensitive, less susceptible to damage."

"You'll find," Vesh-malik said, gently leading him into an adjacent chamber, "that your diet will no longer be so particular. You'll have less need for rest. You will be stronger. In general, your body will be more adaptable, more versatile, less likely to damage or to wear out."

But something was wrong. Satina's clothes didn't fit.

"Is this my uniform?" he asked, trying to fasten his shirt. "This is much too small."

"No, rather your chest has enlarged. All your organs are somewhat larger than before, including your lungs. This is a result of our surgery, though we touched none of your innards directly.

"Your body is reacting to the basic general change, which we made upon you. We did not directly alter any muscle, organ, tissue, or bone. Merely touched the cells in a certain way, made sure that the touch was retained and not canceled by your body's defense mechanisms, then put you back together."

Other science fiction novels and stories featuring characters with special powers include:

- H. Rider Haggard's *She* (1886).
- Alfred Jarry's *The Supermale* (1902).
- Guy Dent's *The Emperor of If* (1926).
- John Russell Fearn's *The Intelligence Gigantic* (1933).
- Olaf Stapledon's *Odd John* (1935).
- Stanley Weinbaum's "The Adaptive Ultimate" (1935).
- H. G. Wells' *Star Begotten* (1937).
- Charles Harness' *The Paradox Men* (1953).
- James Blish's *Jack of Eagles* (1952).
- Theodore Sturgeon's *More Than Human* (1953).
- Wilmar H. Shira's *Children of the Atom* (1953).
- Arthur C. Clarke's *Childhood's End* (1953).
- Wilson Tucker's *Wild Talent* (1954).
- Algis Budrys' "Nobody Bothers Gus" (1955).
- Frank Robinson's *The Power* (1956).
- John Brunner's *The Whole Man* (1964).

In the 1970s television series *The Six Million Dollar Man*, Steve Austin is given super strength and speed through the miracle of bionics—an artificial arm, artificial eye, and two artificial legs. He can leap twenty feet or more into the air, lift cars, and run at speeds up to sixty miles per hour.

But these are all made-up stories: pure fantasies. So why would we include "super-beings" in our list of science fiction predictions that have become science fact?

Because in 2004, *scientists in Germany found a five-year-old child with "super strength"*—the world's first and only known "superman." Through a genetic mutation, the boy has been endowed with bulging arm and leg muscles, and can lift weights other children his age cannot handle. The boy's mutant DNA was found to block production of myostatin, a protein that limits muscle growth. By studying his mutation, the scientists hope to find treatment for patients with atrophied muscles.

The idea of an injection to repair atrophied muscles is the basis of the comic book character Captain America. Steve Rogers, a young man whose arms and legs have been weakened by childhood polio, is injected with a "super soldier serum" which transforms him into a man with peak strength, stamina, and agility.

Supernova

T HE CONCEPT OF NOVAS and supernovas in science fiction is relatively recent; astronomers were way ahead of the sf writers in this instance.

In 1572, Danish astronomer Tycho Brahe saw what he thought was a new star in the night sky. When he first saw it, it was brighter than Venus. But it gradually faded over the next four months until it could no longer be seen at all. Tycho (he is usually known by his first name) described what he saw as *nova stella* (a "new star"), and because of this, stars that suddenly appear in the sky are called novas.

"Of course, novae, as we now know, are not new stars but old ones that have exploded," explains Isaac Asimov in *Words of Science* (1959). "Occasionally, though, a star tears itself apart completely and increases in brightness a few hundred-billionfold. Such a star is a *supernova*."

A supernova occurs at the end of a star's lifetime, when its nuclear fuel is exhausted and it is no longer supported by the release of nuclear energy. If the star is particularly massive, then its core will collapse and in so doing will release a huge amount of energy. This will cause a blast wave that ejects the star's envelope into interstellar space. The result of the collapse may be, in some cases, a rapidly rotating neutron star that can be observed many years later as a radio pulsar.

While many supernovae have been seen in nearby galaxies, they are relatively rare events in our own galaxy. The last to be seen was Kepler's star in 1604. This remnant has been studied by many X-ray astronomy satellites, including ROSAT.

There are, however, many remnants of supernovae explosions in our galaxy. These are seen as X-ray shell-like structures caused by the shock wave propagating out into the interstellar medium.

Another famous remnant is the Crab Nebula, which exploded in 1054. In this case a pulsar is seen which rotates thirty times a second and emits a rotating beam of X-rays (like a lighthouse). Another dramatic supernova remnant is the Cygnus Loop.

Murray Leinster's "First Contact" (1945) features a meeting near the remnants of a supernova between a human and alien ship, who work together to find a compromise that benefits both and prevents mutual destruction. James Gunn used the Crab Nebula—the cloudy remnant of a supernova that was observed on Earth in 1054 by Chinese and Japanese astronomers—at the end of *The Listeners* (1972) to evoke the idea that civilizations go to the trouble of communicating only when faced with total destruction.

In Arthur C. Clarke's "The Star" (1955), a priest with a scientific team exploring a nova discovers that it destroyed an advanced civilization to provide the Star that shone over Bethlehem at the birth of Jesus Christ.

Suspended Animation (see also "Cryogenic Preservation")

FORMS OF LIMITED SUSPENDED ANIMATION occur in nature all the time. We call it *hibernation*. During winter, when the temperatures drop below freezing and food is scarce, a hibernating animal finds a den and goes into a long sleep that can last uninterrupted for several months.

The hibernating animal drops into a state in which its basal metabolism rate becomes greatly reduced. During hibernation, the animal's body temperature may drop to as little as forty degrees Fahrenheit.

Hibernation is an instinctive response that helps the animal survive the harsh winter. In hibernation, the animal does not eat, and instead lives off accumulated body fat. The lower metabolism rate means the animal burns less energy and therefore needs far fewer calories to survive.

Bears and other warm-blooded mammals are not the only creatures that can enter a state of suspended animation. Fish and invertebrates can also enter suspended animation, enabling them to survive long periods of time without oxygen. Humans who have fallen through the ice in a frozen lake have also entered a state of hypothermia; their metabolism slows, enabling them to survive after being underwater and without air for up to twenty minutes.

In one scientific study, scientists have isolated in a species of worm the specific gene that controls the animal's ability to enter suspended animation when deprived of oxygen. In another experiment, scientists induced a state of suspended animation in zebra fish embryos by depriving them of oxygen, which shuts down all observable metabolic activity, including the heartbeat. In this suspended animation state, the embryos survived without oxygen for twenty-four hours. When given oxygen once more, their metabolic rate returns to normal with no harm to the fish.

Cold temperatures have also been used to reduce the metabolic rates of patients during surgery. For instance, in a 1990 operation at Columbia Presbyterian Medical Center, a patient's body temperature was reduced from 98.6 degrees to 72 degrees Fahrenheit. The cooling process stopped all blood flow, and the patient's heart stopped pumping, allowing surgeons to perform a delicate operation on a thin blood vessel pressing dangerously on vital brain centers. The patient lay on the operating table with no heartbeat for nearly half an hour before the body was warmed and heart beat restored. While the brain can only survive a few minutes without oxygen when the body is at normal body temperature, it can survive up to an hour without oxygen when chilled.

In April 2005, researchers at the Fred Hutchinson Cancer Research Center placed mice into a state of suspended animation by exposing them to low concentrations of hydrogen sulfide. Within five minutes of breathing air to which eighty parts per million hydrogen sulfide had been added, the mice's oxygen consumption dropped by half, body temperature fell from 98.6 to as low as 51.8 degrees Fahrenheit, and respiration dropped from 120 to 10 breaths per minute.

The oldest version of suspended animation is when someone falls asleep and wakes up years later, like Washington Irving's "Rip Van Winkle" (1819). Suspended animation was used by William Bellamy in *Looking Backward: 2000–1887* (1888) in which the protagonist, suffering from insomnia, tries hypnotism, is buried in an earthquake, and wakes up more than a century later. Even earlier, Edgar Allen Poe used hypnotism to delay the decay of a corpse in "The Facts in the Case of M. Valdemar" (1845).

The classic story of suspended animation is H. G. Wells' 1899 novel *When the Sleeper Wakes*. The protagonist awakens after two centuries of suspended animation to find everyone he knew is dead and the world of the future is far beyond what he could ever have imagined. But there are perks to his two-hundred-year doze. Through the magic of compound interest, not spending any of his money for 203 years has made him one of the wealthiest and most important people on the planet:

> "You must understand," began Howard abruptly, avoiding Graham's eyes, "that our social order is very complex. A half explanation, a bare unqualified statement would give you false impressions.
>
> "As a matter of fact—it is a case of compound interest partly—your small fortune, and the fortune of your cousin Warming which was left to you—and certain other beginnings—have become very considerable. And in other ways that will be hard for you to understand, you have become a person of significance—of very considerable significance—involved in the world's affairs."

The astronauts in the original *Planet of the Apes* movie (1968) are kept in suspended animation during their long space journey. Placing astronauts in suspended animation for long-distance space travel is also used in the film *2001: A Space Odyssey* (1968).

Other sf works that feature suspended animation as a theme or plot device include:

- A. E. van Vogt's "Far Centaurus" (1944).
- Raymond F. Jones' *The Alien* (1951).
- Robert Heinlein's *The Door into Summer* (1957).
- Larry Niven's *The World of Ptavvs* (1966).
- Michael Moorcock's and Hilary Bailey's *The Black Corridor* (1969).
- James White's *The Dream Millennium* (1974).
- Colin Wilson's *The Space Vampires* (1976).
- Orson Scott Card's *The Worthing Saga* (1990).
- Richard Ben Sapir's *The Far Arena* (1978).
- Richard Lupoff's *Sun's End* (1984).

And in the cartoon *Futurama*, created by Matt Groening, a lonely loser named Fry accidentally falls into a suspended animation chamber. He awakens in the far future and gets a job as a delivery boy for an interplanetary freight service, where he falls in love with Leela, a purple-haired, one-eyed Cyclops.

Several sf writers have imagined that a living organism can be put into a state of suspended animation by converting it into a crystalline or powdered form. Many organisms seem able to live in such a suspended form for decades or even centuries, and then be brought to active life. When I was a child, comic books advertised "live sea monkeys" as mail order pets (these kits are still available today). When you ordered, you received a small container of brine shrimp eggs, resembling granules of brown sugar. By placing the eggs in warm salt water, you could hatch them and raise baby brine ship. I always wondered how long those eggs had lain in a dormant state as powder in a jar. Was it months, years, or even centuries? Other creatures, such as bacteria, can also survive in a spore or dormant state, and then be restored to active life when heat and nutrients are applied.

In an episode of the original *Star Trek*, aliens put certain members of the U.S.S. *Enterprise*'s crew in suspended animation by converting them into pyramid-shaped blocks of crystal.

And in Henry Kuttner's 1940 short story "Beauty and the Beast," a race of reptilian creatures on Venus survives a plague by converting themselves to a crystalline material, in which they remain in suspended animation until exposed to solar radiation:

> The life energy had been drained from one of the reptiles. As the electrons drew in toward their protons, there had been a shrinkage...and a change. A jewel of frozen life, an entity held in absolute stasis, lay before the Venusian scientists, waiting for the heat and solar rays that would waken it to life once more.

Teleportation and Transporters

STAR TREK FIRST INTRODUCED the notion of teleportation—instantaneous transmission of matter between two points—to a mass TV audience. The *Enterprise* and other ships in the Federation all were equipped with transporters, a device used to teleport personnel from the ship to the surface of whatever planet they were visiting.

Teleportation has a long history as an sf notion. In George Langelaan's 1957 short story "The Fly," made into a famous film of the same title, a scientist invents a teleportation device. It works, but a fly accidentally gets into the teleportation chamber with the scientist when he tests the device. When he materializes in the other chamber, he has a fly's head, the fly now wearing his. His famous line from the movie, uttered by the terrified human-headed fly in a high-pitched voice: "Help me! Help me!"

Teleportation devices also play a prominent role in Barry N. Malzberg's novel *Guernica Night* (1974), in which the protagonist is concerned about the effects of teleporting on human health:

> Although uses of the Transporter are theoretically unlimited, it is well-known that there is an absolute number of trips which one may take before a certain deterioration sets in, and also, trips should not be spaced too closely together because of an exhaustion of the cell-reproduction qualities.
>
> I am well under the approved limits of use for one of my age; I do not like to enjamb uses, and there is once again that wrenching terror when I commit myself to the machine, that familiar sensation of irreducible loss. There is a moment in which we do not exist, there is a moment in which we inhabit all space and time, and only the instinctive chemistry of the cells carrying its imprint enables us to become what we were, where we want to be.
>
> That chemistry is unique and irreducible; therefore the Transporter cannot err...but in an infinite universe, there is an infinity of choice

and someone at some time will crawl out of these machines in a different time, a monster.

In A. E. van Vogt's *The World of Null-A* (1945), one of the feats Gilbert Gosseyn masters is the ability to teleport himself anywhere, as long as he has memorized the location down to "seventeen decimal points."

In his classic 1956 novel, *The Stars My Destination*, Alfred Bester describes the process of "jaunting"—the ability to teleport without the aid of a transporter or other mechanical device. Humans could self-teleport to any destination they could mentally picture and which was within their jaunting range; different people had different ranges:

> We have established that the teleportive ability is associated with the Nissi bodies, or Tigroid Substance in nerve cells. The Tigroid Substance is easiest demonstrated by Nissi's method using 3.75 g. of methylene blue and 1.75 g. of Venetian soap dissolved in 1.000 cc. of water. Where the Tigroid Substance does not appear, jaunting is impossible. Teleportation is a Tigroid Function.
>
> Any man was capable of jaunting provided he developed two facilities, visualization and concentration. He had to visualize, completely and precisely, the spot to which he desired to teleport into a single thrust to get him there. Above all he had to have faith. He had to believe he would jaunt. The slightest doubt would block the mind-thrust necessary for teleportation.
>
> The limitations with which every man is born necessarily limited the ability to jaunt. Some could visualize magnificently and set the co-ordinates of their destination with precision, but lacked the power to get there. Others had the power but could not, so to speak, see where they were jaunting. And space set a final limitation, for no man had ever jaunted further than a thousand miles. He could work his way in jaunting jumps over land and water from Nome to Mexico, but no jump could exceed a thousand miles.

Other stories in which teleportation is a psi power include James Gunn's "Wherever You May Be" (1953), James H. Schmitz's *The Witches of Karres* (1966), Tom Reamy's *Blind Voices* (1978), and Timothy Zahn's *A Coming of Age* (1984).

In Stephen King's 1981 short story "The Jaunt," people are rendered unconscious using gas before being jaunted by machine, because time perception is different during jaunting than in the physical world. While the jaunt takes a nanosecond, it seems like eons to the human mind, driving the jaunted person mad unless he or she is asleep. In King's story, a boy holds his breath so he can stay awake and experience the jaunt. At the destination, his father awakens to see the boy screaming, insane, hair turned white, ranting and raving while ripping out his eyes with his hands. "Longer than you think, Dad," the boy cackles insanely, "longer than you think!"

Here are some of the many other sf works that use teleportation as a plot device or theme:

- Edward Page Mitchell's "The Man Without a Body" (1877).
- Robert Duncan Milne's "Professor Vehr's Electrical Experiment" (1885).
- Fred T. Jane's *To Venus in Five Seconds* (1897).
- Garrett P. Serviss' *The Moon Metal* (1900).
- Clement Fezandie's "The Secret of Electrical Transmission" (1922).
- Ralph Milne Farley's *The Radio Man* (1924).
- Edmond Hamilton's "The Moon Menace" (1927).
- Norman Matson's *Doctor Fogg* (1929).
- Jack Williamson's "The Cosmic Express" (1930).
- George O. Smith's *Venus Equilateral* (1947).
- Gordon R. Dickson's *Time to Teleport* (1960).
- Poul Anderson's *The Enemy Stars* (1959).
- Damon Knight's *The People Maker* (1959).
- Clifford Simak's *Way Station* (1963).
- Lloyd Biggle's *All the Colours of Darkness* (1963).
- Philip K. Dick's *The Unteleported Man* (1964).
- Thomas M. Disch's *Echo Round His Bones* (1967).
- Harry Harrison's *One Step from Earth* (1970).
- Joe Haldeman's *Mindbridge* (1976).
- Vernor Vinge's *The Witling* (1976).
- Walter Jon Williams' *Knight Moves* (1985).
- David Langford's and John Grant's *Earthdoom!* (1987).
- Dan Simmons' *Hyperion* (1989).
- Steven Gould's *Jumper* (1992).

Could teleportation actually be a real psi power, along with telekinesis and telepathy, both of which have been investigated for decades? In their book *Science of the X-Men* (2000), Link Yaco and Karen Haber present a theory of how Nightcrawler, a member of the mutant team of the X-Men, is able to teleport without the help of a transporter:

Space is dimensionality, specifically three dimensions, if you're considering our space-time continuum. And space is also related to matter in that, if you have a great deal of matter in one place—like a black hole—space collapses and everything "rolls downhill" into the collapsed space.

Another word for this is gravity. If you consider the planet Earth to be a depression in the fabric of space, it has enough mass to not only bend the fabric of space but to create a lot of gravity as well. Our planet has an equatorial diameter of almost eight thousand miles and most of that is rock and metal. That's enough mass so that it literally bends the fabric of space and creates a lot of gravity.

Any one or thing attempting to escape that gravitational pull must have sufficient momentum and/or the ability to "step" outside regular space. In fact, that's one way to take a shortcut through space: just bend it with a lot of mass.

If he [the teleporting man] doesn't follow the natural curve of three dimensional space, he must to able to go outside of it. The current thinking is that our universe is composed of about twelve dimensions. However, all but four are rolled up into unimaginable thinness.

We can only guess that a dimensional traveler is able to compress himself in a similar manner, without any bodily harm.

The theoretical basis for mechanical teleportation already exists: according to an article in *Scientific American*, the smallest unit of information is a quantum bit, or qubit. Information is not purely mathematical. It always has a physical embodiment, and the unit of quantum information is the qubit.

Two qubits can become "entangled," or connected as a pair, in which case they form an "e-bit." Unlike atomic particles, qubits do not have to be in close proximity to form an e-bit pair; they could theoretically be light years apart. A qubit can be teleported within the e-bit, allowing a quantum bit to instantaneously vanish and disappear at a distant location.

Quantum teleportation is not just theory; it has already happened. One form of teleportation is to duplicate the atomic arrangement of the thing being duplicated at another location. The idea is that if you create an identical duplicate, you have essentially teleported the item, especially if the original is destroyed in the process.

A team of physicists in Denmark used a beam of light to induce, in a group of atoms in a cloud of cesium gas, the same quantum spin of another group of such atoms. In essence, the beam of light teleported a distant cesium gas cloud to the laboratory in Denmark.

Far easier than trying to explain teleportation is simply to let it happen, as Edgar Rice Burroughs did in his John Carter series of novels, in which Captain John Carter teleports between Earth and Mars through sheer willpower. Here's how Burroughs describes it in his novel *The Gods of Mars* (1918):

> With arms outstretched toward the red eye of the great star I stood praying for a return of that strange power which twice had drawn me through the immensity of space...
>
> I turned my eyes again toward Mars, lifted my hands toward his lurid rays, and waited.
>
> Scarce had I turned ere I shot with rapidity of thought into the awful void before me. There was the same instant of unthinkable cold and utter darkness that I had experienced twenty years before, and then I opened my eyes in another world....

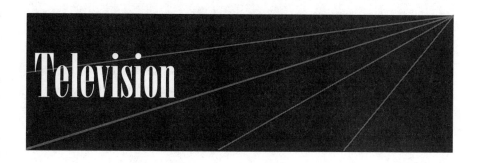

Television

THE MOST SALIENT early description of television is Mark Twain's "From the London Times of 1904" (1898).

Hugo Gernsback predicts television in *Ralph 124C 41+*—originally published as a series of chapters in Gernsback's magazine *Modern Electrics* (1911–1912) and then as a novel in 1925. The title is the name of the hero, Ralph. The numerals 124C 41 in his name refer to a scientific designation, and the "+" is an honorific for which only half a dozen people in Ralph's era are qualified to add their name. The novel is mostly a framework for a series of predictions about future technology and how life will be changed by them. In addition to predicting television, the novel also predicts clothing made of glass, night baseball, aquacades, and radar, the last one complete with diagrams.

As a matter of fact, Gernsback was a pioneer in television, back when there were no receivers, as he was in radio, when he sold his "Telimco Wireless" with a receiver and a sender, since there were no radio stations on the air.

In 1906, a Russian engineer, Boris Rosing, produced crude television pictures using a cathode ray tube and an optical scanning system incorporating a rotating disc. U.S. inventor Philo Farnsworth conceived the basic operating principles of electronic television in 1919, at age thirteen, and had built the first practical TV set by 1928.

Although best known as the inventor of TV, Farnsworth had many other scientific accomplishments. He invented the first electron microscope and the first infant incubator. He was also involved in the development of radar and nuclear fusion power. At the time of his death in 1971, he held three hundred U.S. and foreign patents.

How does a modern TV work? A decoder box receives the cable or broadcast signal and splits it into three colors: red, green, and blue. Electron guns fire three beams of electrons through vertical and horizontal deflection coils onto the inside surface of the tube, which serves as the TV screen.

The tube is coated with about a million dots. A third of the dots glow red when bombarded with the electron beams, a third glow blue, and a third glow green. Combinations of these three basic colors make up the color picture you see on your TV screen. The electron beams scan hundreds of lines on the screen to make the picture move.

Terraforming

A MORE RADICAL METHOD of colonizing a planet than building shelters to protect the colonists from the planet's environment is to modify that environment to make it more hospitable to human life—a process called "terraforming."

Jack Williamson invented the term "terraforming" in his stories collected in *Seetee Ship* (1942). Arthur C. Clarke's *The Sands of Mars* (1951) and Patrick Moore's *Mission to Mars* (1955) incorporate minor modifications to the Mars environment.

Carl Sagan popularized terraforming in *The Cosmic Connection* (1973) and Adrian Berry in *The Next Ten Thousand Years* (1974). Isaac Asimov used terraforming in "The Martian Way" (1952) and Poul Anderson terraformed Venus in "The Big Rain" (1954). Pamela Sargent's *Venus of Dreams* (1986) is also built around the idea of terraforming Venus.

Other novels and stories deal with terraforming Mars, particularly Kim Stanley Robinson's trilogy that began with *Red Mars* (1992) and was followed by *Green Mars* (1993), and *Blue Mars* (1996). Robert A. Heinlein's *Farmer in the Sky* (1950) and Poul Anderson's *The Snows of Ganymede* (1958) deal with the terraforming of Ganymede.

In Roger Zelazny's 1969 novel *Isle of the Dead*, Francis Sandow is a professional terraformer whose technology is augmented by godlike powers he has acquired. Terraforming is a theme in other Zelazny stories as well.

In the Arnold Schwarzenegger film *Total Recall*, based on a Philip K. Dick short story, aliens—for reasons never fully explained—have left a giant terraforming device on Mars. When Schwarzenegger switches the device on, the machine sends oxygen spewing into the Martian atmosphere, terraforming Mars into a planet on which human beings can live without space suits or controlled-atmosphere cities.

How could you terraform a cold and dry planet like Mars to be more Earth-like? One possible solution: spread dark substances over the polar caps. The dark colors absorb more heat, melting the ice to form rivers and lakes, perhaps even seas.

Another proposed terraforming technique: orbit the planet with huge reflectors to focus sunlight on its surface, thereby raising the temperature. Large plants could also be established to generate and release nitrogen and oxygen into the atmosphere, creating breathable air. Several technologies for building such plants are routinely used to make these gases for industrial applications today.

In one such oxygen-generating system, called "vacuum sleeve absorption" (VSA), a zeolite molecular sieve adsorbent removes nitrogen, carbon dioxide, and water from a stream of compressed air. When the sieve is saturated, a vacuum pump removes the gases, enabling the sieve to be reused. Because the system operates at vacuum, it makes more efficient use of the sieve, which in turn increases overall system efficiency.

VSA plants typically contain dual adsorbent beds, with one bed producing oxygen while the other regenerates, so the system can operate uninterrupted. The two-bed design is simple yet effective, ensuring high system availability while reducing the cost of oxygen production.

Terraforming may soon become a reality—on Earth: in 2005, a Swiss ski resort announced it would combat global warming by wrapping its mountain glaciers in aluminum foil to keep them from melting. And NASA is offering a $250,000 reward to the first researcher to develop a method of generating breathable oxygen from lunar soil. The loose soil covering the Moon's surface is forty-five percent oxygen, but it is bound with titanium and other minerals in a way that makes extraction difficult.

Test-Tube Babies

THE EARLIEST NOVEL to describe test-tube babies was Aldous Huxley's *Brave New World* (1932), in which multiple births (that we would now call clones) were grown to term in a laboratory and then "decanted."

In James Blish's *A Case of Conscience* (1958), a Lithian embryo is saved from the destruction of the planet Lithia. The infant is raised by its Jesuit priest foster-parent and becomes a kind of anti-Christ.

Science fiction and prediction turned into science fact when, on July 25, 1978, Louise Brown, the world's first "test-tube baby"—conceived through the then-new process of in vitro fertilization (IVF)—was born.

Today, one out of six couples is infertile, and an entire "infertility industry" has been created to help these people start families. In 2000, American women underwent about one hundred thousand fertility treatments.

"Assisted Reproductive Technologies" (ART) have been used in the United States since 1981. More than seventy thousand children have been born in the U.S. as a result of ART, including forty-five thousand as a result of IVF. ART works best when the woman's uterus is healthy, her eggs retrievable, and the man has healthy sperm. If not, donor eggs and sperm can be used.

"Artificial insemination" is an umbrella term describing IVF and any other process in which sperm are collected, washed free of seminal fluid, and then placed inside the woman's reproductive organs to fertilize her egg naturally.

The IVF procedure involves introducing a fertilized egg into the uterus. This can be done with eggs harvested from the woman, and her husband's sperm, or donor gametes. Fertilization takes place in a Petri dish in a laboratory, thus the commonly-used terms *test-tube baby* or *test-tube fertilization*.

Another common infertility treatment, gamete infra-fallopian transfer (GIFT), requires the female's egg to be surgically removed from the ovary, combined with sperm, and immediately placed into the fallopian tube, in hopes of fertilization. The term *gamete* refers to the male and female reproductive cells—the egg and sperm.

Men and women routinely freeze sperm and eggs for future conception. A man having prostate surgery or a vasectomy, for example, might freeze his sperm in case the operation renders him infertile. IVF often produces more embryos than are immediately needed, and the extras are typically frozen for possible use later.

Today there are four hundred thousand frozen embryos stored in fertility clinics throughout the United States. Each is a potential adult human being, frozen in cryogenic suspended animation until the parents who "own" the embryo and pay their "rent" (cryogenic storage costs for an embryo are about $1,500 per year) decides whether he or she will be born, discarded, or kept in suspended animation indefinitely.

While IVF was believed for many years to be perfectly safe, a recent study shows that babies born through IVF are two-and-a-half times more likely to have low birth weight (5.5 pounds or lighter) than babies conceived naturally. Another study found that IVF babies are twice as likely as naturally conceived babies to have multiple major birth defects, especially chromosomal and musculoskeletal abnormalities.

Why? One cause might be the drugs used to induce ovulation and maintain the pregnancy in its early stages. Researchers also speculate that IVF may allow a flawed or damaged sperm to penetrate the egg—one that would not otherwise be able to do so under its own power.

Time Travel

HEN DID THE NOTION of time travel begin in sf? Perhaps the most famous sf novel is *The Time Machine* (1895) by H. G. Wells. But fourteen years earlier, Edward Page Mitchell published his time travel short story, "The Clock that Went Backward." In Edward Bellamy's novel *Looking Backward: 2000–1887* (1888), the hero thinks he has traveled back in time, but then wakes up to discover it was a dream. Dreams of time travel were also used in Charles Dickens' "A Christmas Carol" (1843) and Edgar Allen Poe's "A Tale of the Ragged Mountains" (1844).

In Wells' classic novel, the logical explanation for time travel is that we can travel freely within the three dimensions of space; hence, we should be able to do the same within the fourth dimension of time:

> Any real body must have extension in four directions: it must have Length, Breadth, Thickness, and Duration. There are really four dimensions, three which we call the three planes of Space, and a fourth Time. There is no difference between Time and any of the three dimensions of Space except that our consciousness moves along it.
>
> You are wrong to say that we cannot move about in Time. For instance, if I am recalling an incident very vividly I go back to the instant of its occurrence: I become absent-minded, as you say. I jump back for a moment. Of course we have no means of staying back for any length of Time, any more than a savage or an animal has of staying six feet above the ground.
>
> But a civilized man is better off than the average in this respect. He can go up against gravitation in a balloon, and why should he not hope that ultimately he may be able to stop or accelerate his drift along the Time-Dimension, or even turn about and travel the other way?

After Wells, many sf writers used time machines, including William Wallace Cook in *A Round Trip to the Year 2000* (1903). In the novel *The*

Fury out of Time (1965), Lloyd Biggle, Jr., describes the effects of time travel on the time traveler:

Silence. Then almost imperceptibly, pressure. Twisting in alarm, Karvel moved his hands across his chest and touched his face.

There was nothing to brush away, nothing tangible to fight. The pressure continued—light, insistent, all-embracing.

He was moving through time, and time resisted his passage. And there was pressure—feathery, intangible, but nonetheless relentless.

He lay quietly, and suppressed an urge to escape from the cylinder, to find out what was happening outside. The limping, time-compressed minutes slipped away, and slowly, tediously, the pressure increased.

Karvel began to wonder about the meaning of time when one was passing through time. Was his wristwatch actually marking off seconds and minutes and hours?

He lay in the tightening grip of time and pondered its measurement. When finally he decided to perform an experiment, he found the luminous hands of his watch immobilized under a pressure-warped crystal.

Keith Laumer, in his 1971 novel *Dinosaur Beach*, gives his own description of time travel:

I was doubled back on my own time track. The fact that this was a violation of every natural law governing time travel was only a minor aspect of the situation, grossly outweighed by this evidence that the past that Nexx Central had painfully rebuilt to eliminate the disastrous effects of Old Era time meddling was coming unstuck.

And if one piece of the new mosaic that was being so carefully assembled was coming unglued, then everything that had been built on it was likewise on the skids, ready to slide down and let the whole complex and artificial structure collapse in a heap of temporal rubble that neither Nexx Central nor anyone else would be able to salvage.

Technology and psi powers are not the only means by which sf writers have sent their characters through time. Many stories have been written in which time travel is achieved through an accident, like Mark Twain's *A Connecticut Yankee in King Arthur's Court* (1889) and L. Sprague de Camp's *Lest Darkness Fall* (1939). Time travel can also be achieved through magic, as in J. L. Balderston's and J. C. Squire's *Berkeley Square* (1929), Robert Nathan's *Portrait of Jennie* (1940), and Jack Finney's classic *Time and Again* (1970).

Other sf novels and stories about time travel include:

- Ray Cummings' *The Man Who Mastered Time* (1924).
- Ralph Milne Farley's "The Time Traveller" (1931).
- John Taine's *The Time Stream* (1931).
- John Russell Fearn's *Liners of Time* (1935).
- Jack Williamson's *The Legion of Time* (1938).
- Manly Wade Wellman's "Giants from Eternity" (1939).
- Robert Heinlein's "By His Bootstraps" (1941).
- Henry Kuttner and C. L. Moore's "Vintage Season" (1946).
- T. L. Sherred's "E for Effort" (1947).
- Ray Bradbury's "A Sound of Thunder" (1952).
- L. Sprague de Camp's "A Gun for Dinosaur" (1956).
- Ian Watson's "The Very Slow Time Machine" (1978).
- David Lake's *The Man Who Loved Morlocks* (1981).
- Connie Willis' "Fire Watch" (1982).
- David Gerrold's *The Man Who Folded Himself* (1973).

Can the time machines of Wells, Biggle, Laumer, and dozens of other writers actually be built? Paul Davies addresses this question in his book *How to Build a Time Machine* (2001), in which he proposes using a wormhole (see "Wormholes") as the basis for time travel:

To turn a wormhole into a time machine you have to establish a permanent time difference between the two ends. The simplest technique is to use the ordinary time dilation effect.

To do this, the wormhole is given an electric charge (for example, by firing electrons into it) when it is still quite small—say, the size of a subatomic particle. One mouth of the wormhole is then fed into an ordinary circular particle accelerator and whirled around at very near the speed of light, while the other is held still.

This produces a growing temporal discrepancy between the two mouths of the wormhole. The process is allowed to continue for, say, ten years, at which time the moving mouth is brought to rest and allowed to approach the other wormhole mouth.

The wormhole is now able to send particles of matter back in time for up to ten years. In the final step of the process, the wormhole is returned to the inflator factory to be expanded to a size large enough for a human being to traverse—say, ten meters in diameter. Meanwhile, the length of the wormhole is kept as short as possible.

There are a number of technical issues that put the feasibility of such a wormhole time machine project in question, most notably that no one has yet seen a wormhole, much less gotten near one, gone through one, or learned how to control one.

Time travel through a man-made corridor, not a wormhole, was featured in the 1966 television series *The Time Tunnel*. Two scientists, Dr. Tony Newman and Dr. Douglas Phillips, are working on Project TicToc, a government experiment to develop time travel. They jump through the portal before it is fully tested, and spend each episode going to a different time period where of course, they have an adventure.

The 1989–1993 television series *Quantum Leap* also employed a time traveler, Dr. Sam Beckett, played by Scott Bakula—the captain of the U.S.S. *Enterprise* in the *Star Trek* TV series *Enterprise*—who is also constantly jumping through time, caught in a time machine of his own invention. Admiral Albert "Al" Calavicci, a colleague from our time period, can communicate with Dr. Beckett, to whom he appears in the different time periods as a holographic image.

To maintain the proper balance of matter in the space-time continuum, whenever Sam travels into the past, he switches places with another person, who travels forward and is held within the time machine's chamber. However, to people in the time period to which Sam has traveled, he appears as the person he has switched places with, not himself—which of course makes little sense.

Physicist Amos Ori proposes that a time machine could be built by forming a loop that causes a region of space-time to curve back upon itself.

According to Ori's theory, the loop would have to be formed within an empty, doughnut-shaped region of space that is contained within a sphere of matter. The distortion of space-time within the center of the doughnut could be produced by gravity waves or nearby massive objects such as black holes.

To travel back in time, you'd get into a rocket ship and fly around the donut, causing you to go back further and further into the past with each orbit.

Transmutation of Metals

FOR OVER FOUR THOUSAND YEARS, gold has been valued for its beauty and scarcity. It is recognized and valued everywhere in the world and can be turned into cash in virtually every nation on earth.

Gold is indeed rare. If all the gold on the planet were melted down and poured into one giant cube, it would measure only eighteen yards across and weigh just ninety-one thousand tons. That's about equal to the total amount of steel made around the world *every half-hour.*

Ancient alchemists labored in vain in their attempt to turn base metals, like lead, into precious metals, like gold. As Isaac Asimov explains in his book *A Short History of Chemistry* (1965),

> Many chemists throughout the centuries have honestly striven to find the technique for producing gold. Some, however, undoubtedly found it much easier and far more profitable merely to pretend to find the technique and to trade on the power and reputation this gave them.
>
> Bolos, in his writings, apparently gave the details of techniques of making gold, but this may not actually have represented fakery. It is possible to alloy copper with the metal zinc, for instance, to form brass, which has the yellow color of gold. It is quite likely that preparation of a gold-colored metal would be equivalent, to some of the ancient workers, of forming gold itself.

A number of the alchemists wrote books in which they gave partial recipes for transmuting base metals and other materials into gold, but their formulas were neither clear nor complete. In his book *The Story of Chemistry* (1959), Georg Lockemann writes:

> [For] the artificial production of gold, the right mixture of the four elements is required, and the "spirit" (the heated volatile mercury) must penetrate into the "bodies" (copper, lead, etc.), which is facilitated by adding some actual gold.

The mysterious "elixir" itself is obtained only by the correct union of the four elements, the body (metals), the spirit (mercury), the male and the female.

It assimilates the "bodies" and colors them (hence its name "tincture") by transforming them into silver and gold, up to a thousandfold quantity. It gives life to the "bodies" and resurrects the dead.

The "bodies" mean the metals, except mercury, which always contain the four elements—two in an open state and two in a hidden one. To these "bodies" mercury must be added as "spirit."

In the legend of Midas, King Midas wishes for more gold so he can become wealthier and more powerful. He is granted this wish, and everything he touches turns to gold, which is where the expression "the Midas touch" originates. Unfortunately, everything means everything: Midas can no longer feed himself, because the utensils and food he touches immediately turn to gold and therefore become inedible. Eventually his daughter touches him and she, too, turns to gold, much to his sorrow, and he is taught a costly lesson about greed.

Honore de Balzac's *The Quest of the Absolute* (1834) concerns transmutation of metals, and Edgar Allen Poe's "Von Kempelen and His Discovery" (1849) describes the transmutation of lead into gold.

The 1950s television series *Superman*, starring George Reeves, featured an episode in which gangsters force a professor to give them a machine he has built to transmute other metals to gold. It turns out be worthless, not because it doesn't work—it does—but because the metal it converts into gold is platinum, which is even more costly.

DC Comics, publisher of the *Superman* comic books, created a villain with transmutation powers: Element King. He emits a ray from his fingertips that can turn any element into any other element, and forms, with other super villains, a Legion of Super Villains to perpetrate crimes—never apparently realizing he can attain all the wealth he would ever desire just by turning rocks, dirt, and mud into a limitless supply of gold.

While no one has yet invented a machine to turn lead into gold, transmutation is now a reality, thanks to the discovery of nuclear fusion and fission. Every day, atomic scientists routinely turn an atom of one element into an atom of an entirely different element by bombarding it with neutrons or other subatomic particles. For instance, bombarding lithium with neutrons converts it into tritium.

UFOs
(see also "First Contact")

FIRST CONTACT WITH AN ALIEN SPECIES can happen in one of two ways: either we go to them, or they come to us. In the first scenario—we go to them—a manned outer space mission encounters alien life-forms, either by finding them on another planet (either within our solar system or outside) or by crossing paths with an alien spaceship during a space voyage. In the second scenario—they come to us—we make first contact when aliens come to Earth. Because these aliens have mastered interplanetary space travel and we haven't, they would naturally be more technologically advanced.

Their arrival is typically signaled by sightings of Unidentified Flying Objects, or UFOs.

The aliens visiting Earth could be either friendly, as in the Stephen Spielberg movie *E.T.*, or hostile, as in Orson Welles' radio broadcast of *The War of the Worlds*, in which Earth is invaded by Martians.

In *The War of the Worlds*, Earth's military is unable to stop the technologically superior Martian attackers, whose main weapon is some sort of energy beam weapon. We survive by luck: Martian immune systems have no defense against Earthly bacteria, and the Martians are killed off by disease.

In *Signs*, Mel Gibson plays a rural farmer and ex-minister on whose cornfield appear crop circles. It turns out that the crop circles are indeed made by hostile extraterrestrials. They are defeated not by disease but by moisture: water burns them like sulfuric acid burns us. Aliens were also vulnerable to water in the TV series *Alien Nation*, which made its debut in 1989, but it was salt water that was acidic to them; fresh water was fine.

Alien encounters are a major sf theme; one could fill a book with all the references of science fiction novels, stories, and movies involving UFOs. Some of the more notable examples include:

- Theodore Sturgeon's "Mewhu's Jet" (1946).
- George Adamski's *Pioneers of Space* (1949).

- C. M. Kornbluth's "Silly Season" (1950).
- Robert A. Heinlein's *The Puppet Masters* (1951).
- Chad Oliver's *Shadows in the Sun* (1954).
- Gore Vidal's *Messiah* (1954).
- Dennis Wheatley's *Star of III-Omen* (1952).
- Stan Layne's *I Doubted Flying Saucers* (1958).
- Raymond Fowler's *The Melchizedek Connection* (1961).
- Fritz Leiber's *The Wanderer* (1964).
- Larry Maddock's *The Flying Saucer Gambit* (1966).
- George Earley's *Encounters With Aliens* (1968).
- Marian Place's *Brad's Flying Saucer* (1969).
- Martin Caidin's *The Mendelov Conspiracy* (1969).
- Keo Felker Lazarus' *The Gismo* (1971).
- Patrick Tilley's *Fade-Out* (1975).
- George H. Leonard's *Alien* (1977).
- Alan Dean Foster's *Alien* (1979).
- Jacques Vallée's *Alintel* (1986).
- Whitley Strieber's *Majestic* (1989).
- David Bischoff's *Abduction: The UFO Conspiracy* (1990).

Real UFOs, though not identified as such, were reported by Charles Fort in his collection of newspaper accounts of inexplicable phenomena, beginning with *The Book of the Damned* (1919). There was a rash of UFO sightings by passengers of ships crossing the Atlantic in the winter and spring of 1896–1897, although these were spaceship-shaped rather than saucer-shaped.

The term "flying saucer" was coined by Kenneth Arnold in 1947 after a sighting near Mount Ranier in Washington state and was popularized by Donald E. Keyhoe in his book *The Flying Saucers Are Real* (1950) and by George Adamski and Desmond Leslie in *Flying Saucers Have Landed* (1953). Conspiracy theorists hold that the federal government is covering up the 1947 crash-landing of an alien spaceship near Roswell, New Mexico, in which parts of the ship, alien bodies, and one live alien (who lived only a short time) were allegedly recovered.

Fort influenced later sf writers and editors—including John Campbell, Damon Knight, and Eric Frank Russell—partly through the reprinting of portions of his books in science fiction magazines in the 1930s. The Fortean society, for those who believe UFOs are real, is still in existence.

There have been hundreds if not thousands of UFO sightings reported throughout a good part of the twentieth century. As Robert Sheaffer writes in his book *UFO Sightings: The Evidence* (1998):

> Flying Saucers, also known as Unidentified Flying Objects or UFOs, have been in the news more or less continuously since 1947, when the famous June 24 sighting by pilot Kenneth Arnold, alleging that a formation of strange disks was zipping up and down Washington's Mt. Ranier, burst into the headlines.
>
> There are many who claim that UFOs can only be explained as extraterrestrial or even more exotic phenomena: interpenetrating universes, "alternate realities," and the like.
>
> There are many people who claim to have actually met the alleged occupants of flying saucers and even to have taken rides in the craft. The late George Adamski was the best known of these so-called contactees: he claimed to have held lengthy discussions with people from Venus.
>
> Others claim to have been "kidnapped" or abducted aboard a strange craft by alien beings, where they were subjected to a bizarre medical-type examination, and afterward set free with their conscious memories of the incident erased.
>
> There are many reports on file of "close encounters" with a mysterious craft, whose description seems to rule out any known object. Maneuvers are often reported that would be plainly impossible for any device of human construction.

Skeptics ask this logical question of UFO believers: "If aliens from an advanced civilization have been visiting Earth in spaceships for decades, how come none of them has openly reached out to us by now?" And until E.T. appears on the evening news and commands the world to "Take me to your leader," the skeptics are likely to remain so.

Undersea Cities

MOST OF US ARE FAMILIAR with the legend of Atlantis, the undersea city, which is rumored to have sunk 11,600 years ago at the end of the Pleistocene Ice Age. Plato, the Greek philosopher, wrote a detailed account of Atlantis as being the home of a great civilization skilled in architecture and engineering. According to Plato, the people of Atlantis became complacent and their leaders arrogant. In punishment, the gods destroyed, flooded, and submerged the city in a single day and night.

In the 1936 film *Undersea Kingdom,* Atlantis is a scientifically advanced undersea civilization planning to invade the surface world with ray guns and robots called "Volkites." The psychic Edgar Cayce predicted that Atlantis would emerge from the ocean in 1969, but it never happened.

In *Mathematical Magic* (1648) John Wilkins wrote about submarines and the possibilities of undersea colonization. Atlantis was featured in a number of undersea stories such as Andre Laurie's *The Crystal City Under the Sea* (1896), Arthur Conan Doyle's *The Maracot Deep* (1929), Stanton Coblentz's *The Sunken World* (1928), and Dennis Wheatley's *They Found Atlantis* (1936).

Jack Williamson's *The Green Girl* (1930) takes place partly under the ocean, and his and Frederik Pohl's *Undersea Quest* (1954) deals with undersea colonization. Undersea life on other worlds is represented by Neil R. Jones' "Into the Hydrosphere" (1933), Henry Kuttner's and C. L. Moore's "Clash by Night" (1943), and James Blish's "Surface Tension" (1952).

The General Motors "Futurama" exhibit at the 1964 World's Fair featured a realistic scale model of what an underwater city might look like. As I recall, it looked much like any other city, though it was enclosed in a plastic dome. Will a real undersea city ever exist? Captain Nemo, the protagonist of the 1869 Jules Verne novel *20,000 Leagues Under the Sea,* says, "I can imagine the foundations of nautical towns, clusters of submarine houses, which, like the *Nautilus*, would ascend every morning to breathe at the surface of the water, free towns, independent cities."

In his novella *The Eve of Rumoko* (1969), Roger Zelazny writes about an undersea city endangered by the concussive force of blasting and drilling a deep hole at the bottom of the ocean where the Earth's crust is thinnest.

In Henry Kuttner's 1956 novel *Destination Infinity*, Earth has been destroyed and humanity has relocated to Venus. While the surface is uninhabitable, the remaining population of humans lives in the oceans of Venus in dome-shaped undersea cities.

Science fiction usually pictures undersea cities as existing under a glass dome or bubble; the residents keep dry and breathe air. Some writers, however, seek an alternate method of surviving underwater. Alexander Beliaev's *The Amphibian* (1928) and James Blish's "Sunken Universe" (1942) both describe people transformed into creatures capable of living under water. Norman Knight's *A Torrent of Faces* (1967) engineers people into "tritons" to live under water.

Science has long been exploring the feasibility of humans breathing water. As Lois H. Gresh and Robert Weinberg explain in their book *The Science of Superheroes* (2002),

> In the 1960s, Dr. J. Kylstra, a scientist at SUNY Buffalo, discovered that saline solutions could be saturated with oxygen gas at high pressure. Kylstra performed experiments on mice to see if they could move the saline solution in and out of their lungs while extracting enough oxygen to survive. The mice could breathe the liquid.
>
> Following Kylstra's lead in 1966, Dr. Leland Clark substituted fluorocarbon liquids for saline solution, as both oxygen and carbon dioxide were both soluble in such liquids. Dr. Clark reasoned that animals could absorb the oxygen from the fluid and replace it with carbon dioxide.
>
> The lower temperature of the fluorocarbons was directly related to how long the mice could survive in fluid. The colder the fluid, the slower the mice breathed, which prevented a buildup of carbon dioxide.
>
> Experiments using fluorocarbons continued throughout the 1990s with increasing success. Tests were performed on dogs without causing major damage to their systems.
>
> One of the problems working with fluorocarbons was that human tissue retained the fluorocarbons. The creation by Alliance Pharma-

ceuticals of perfluoron, fluorocarbon not absorbed by the body, finally made it possible to use liquid breathing in human medical procedures.

In James Cameron's 1989 movie *The Abyss*, characters use an advanced form of fluid breathing to dive to incredible depths. It's science fiction in the movie, but obviously could someday become fact. From there it is not a radical step to breathing water.

Genetic engineering, mutation, and surgery are other alternatives for populating an undersea city. In the 1995 film *Waterworld*, Kevin Costner is a mutant whose gills enable him to breathe underwater. In Richard Setlowe's 1980 novel *The Experiment*, a man with lung cancer has his lungs removed and is surgically given artificial gills which enable him to live as long as he stays submerged:

General Electric developed the thin silicone membrane in the early sixties. Bodell, a researcher in thoracic and cardiovascular surgery in Chicago, described what he called an artificial gill in 1965 using Silastic capillary tubing developed by Dow Corning. It was a device to oxygenate the habitats of rats underwater.

In 1966, just a year after Bodell published, Ayes was awarded a patent on a gill-type underwater breathing device. However, the major medical breakthroughs have come about in just the past decade. In open-heart surgery, both the heart and lungs are bypassed. The blood is totally oxygenated outside the body. Efficient artificial gill-like devices about a liter in size are now standard in heart-lung machines.

My research has been directed at getting back to biological basis. Using the same silicone materials and surgical techniques, I've developed an artificial gill that can be implanted into mammals. It will sustain them underwater for prolonged periods.

A water-breathing gilled man was also featured in the short-lived 1977 television series *The Man from Atlantis*. The water breather was played by Patrick Duffy, best known as Bobby Ewing from the television series *Dallas*.

Numerous sf stories deal with some aspect of underwater living, whether deep ocean exploration or establishing communities on the ocean floor. These include:

- Arthur C. Clarke's *The Deep Range* (1954).
- Jack Williamson's and Frederik Pohl's *Undersea Quest* (1954).

- Kenneth Bulmer's *City Under the Sea* (1957).
- Gordon R. Dickson's *The Space Swimmers* (1967).
- Hal Clement's *Ocean on Top* (1973).
- Lee Hoffman's *The Caves of Karst* (1969).
- Kobe Abé's *Inter Ice Age 4* (1959).
- T. J. Bass' *The Godwhale* (1974).

You may be able to live in an underwater city, at least temporarily, sooner than you think. An underwater hotel, the Hydropolis, is being built, at a cost of half a billion dollars, on the floor of the Persian Gulf sixty-six feet below the surface. Rooms and recreation areas will have bubble walls and domes so guests can see the water at all times.

Virtual Reality

A NUMBER OF STORIES and novels have dealt with people being drawn into fictional worlds, such as Lewis Carroll's *Through the Looking Glass* (1871) and A. Merritt's *The Ship of Ishtar* (1924). In L. Ron Hubbard's *Typewriter in the Sky* (1951), an author falls into his own pirate story.

These are equivalent to virtual reality, but today the term "virtual reality" refers to a simulated or artificial reality created using a computer. The term is credited to Jason Lanier, founder of VPL Research Inc., and was applied to the gloves used as controllers in computer simulations and games, and to the helmets that make the computer world seem real.

In Ray Bradbury's short story "The Veldt," published in his 1951 book *The Illustrated Man,* families can have built onto their houses special projection rooms that create virtual reality environments for recreation. The virtual reality becomes a little too real, and the homeowners in the story are eaten by lions when their children program the virtual reality room for an African jungle.

Ben Bova, in his 1963 story "The Perfect Warrior," describes a virtual reality machine that works by manipulating brain waves: "[it] is nothing more than a psychonic device [that] allows two men to share a dream world created by one of them."

Roger Zelazny's 1997 virtual reality novel *Donnerjack* features a computer-generated virtual reality universe called Virtu. The protagonist, Donnerjack, is considered something of a god in Virtu (though far from the most powerful god there) because he helped create it. People who live in the real world, called Verite, can plug into and enter Virtu in much the same way as Neo can plug into and enter the virtual reality universe of *The Matrix.*

The idea of a matrix was invented by sf writer William Gibson. His Neuromancer trilogy (1984–1988) popularized the concept in stories in which hackers with plugs in their skulls "jacked" into a computer

reality, something Gibson described as projecting "a disembodied consciousness into the consensual hallucination that was the matrix."
But there were predecessors, such as:

- Arthur C. Clarke's *Against the Fall of Night* (1953).
- James Gunn's *The Joy Makers* (1961).
- Daniel Galouye's *Counterfeit World* (1964).
- Larry Niven's and Steve Barnes' Dream Park series (1981–1991).

Other virtual reality novels include:

- Hugh Walker's *War-Gamers World* (1978).
- Vernor Vinge's *True Names* (1981).
- Fred Saberhagen's *Octagon* (1981).
- Gillian Rubinstein's *Space Demons* (1986).
- Andrew Greeley's *God Game* (1986).
- Kim Newman's *The Night Mayor* (1989).

In the 1982 science fiction film *Tron*, a computer programmer becomes trapped in a virtual reality world of sorts: he is split into molecules and transported inside his computer. In this computer-based virtual reality, an evil program called Master Control, similar in demeanor to Agent Smith from *The Matrix*, rules with an iron hand. The computer programmer enlists the aid of other programs, including a security program called TRON, and together they defeat the Master Control. Other virtual reality films include *Welcome to Blood City* (1977), *Brainstorm* (1983), *Dreamscape* (1984), *The Lawnmower Man* (1992), and *Dark City* (1998).

In his 1974 novel *The Eden Cycle*, Raymond Gallun describes what could be viewed as the predecessor to the Matrix—a device called the Sensory Experience Simulator (SES). Like the Matrix, the SES feeds sensory impressions directly into human brains so people can experience anything they desire without any of it being real. Like Neo and Morpheus in *The Matrix*, the protagonists of Gallun's novel, Joe and Jennie, reach the opinion that the SES is too false and unreal.

In the Matrix trilogy of motion pictures, virtual reality takes an evil twist. A superintelligent computer rebels against its human makers, instigating a war that ends with the humans setting off the world's nuclear arsenal, killing most of the population. As the surface has been made uninhabitable because of radiation and nuclear winter, the survivors are forced to live in underground cities warmed by the heat of the Earth's core. Because nuclear winter blots out the Sun, the solar cells that power the supercomputer are running down. As Lyle Zynda explains in his essay in the book *Taking the Red Pill: Science, Philosophy and Religion in The Matrix* (2003),

> In the Matrix, most of humankind is used as a source of power by highly intelligent machines, centuries in the future. Humans are placed from birth in a dreamlike state, in which a world like ours is simulated for their sleeping minds. The machines know that our sense organs convert information from the world (light, sound, etc.) into electrical signals, which are then processed by the brain into the image of reality that constitutes our conscious experience. So, they feed the same electrical signals into the brains of humans that a real world would, creating an illusion indistinguishable from reality.

You could argue that the Matrix is a metaphor for television, and that the average American spends hours a day engaged motionless in such a virtual reality, living life as portrayed on TV images. In much the same way, today's popular MMRGP (massively multiplayer role-playing games) provide a highly addictive virtual reality in which gaming enthusiasts can stay immersed for hours on end.

With the advances being made in computer simulations and software, virtual reality is becoming real. Pilots now train on flight simulators that create the sensation of actually being at the controls in the cockpit. "Four dimensional" theaters show short films that come close to creating a virtual reality. Not only is the picture shown in 3D and the sound heard in stereo, but your chair shakes to simulate the rumble of an earthquake or dinosaur stampede, and when the dinosaur turns toward the audience and sneezes, a fine mist of water is shot into your face out of the back of the chair in front of you.

As technology improves, virtual reality moves closer and closer to a near-perfect duplication of reality. Ray Kurzweil describes it in his book *The Age of Spiritual Machines* (1999):

> Once we are in a virtual reality environment, our own bodies—at least the virtual versions—can change as well. We can become a more attractive version of ourselves, a hideous beast, or any creature real or imagined as we interact with the other inhabitants in each virtual world we enter.

Virtual reality is not a (virtual) place you need go to alone. You can interact with your friends there (who would be in other virtual reality booths, which may be geographically remote). You will have plenty of simulated companions to choose from as well.

So far, it all sounds like fun and games—no more harmful than Game Boy or XBox. But then what Kurzweil predicts comes dangerously close to *Matrix* territory:

Later in the twenty-first century, we will be able to create and interact with virtual environments without having to enter a virtual reality booth. Your neural implants will provide the simulated sensory inputs of the virtual environment—and your virtual body—directly in your brain.

Ultimately, your experience would be highly realistic, just like being in a real world. More than one person could enter a virtual environment and interact with each other. In the virtual world, you will meet other real people and simulated people—eventually, there won't be much difference.

In the second half of the twenty-first century, a typical "Web site" will be a perceived virtual environment, with no external hardware required. You "go there" by mentally selecting the site and then entering that world. Of course, there may be a small charge.

Researchers at Washington State University's Human Interface Technology (HIT) Laboratory are working on technologies that may help humans directly connect with both virtual realities and the computers that generate them. One of their innovations, the Virtual Retinal Display (VRD), scans images directly onto the retina. Another, the MagicBook, enables a reader wearing a special pair of goggles to focus an image on a flat screen; when the user presses a button, a 3D object pops up. Many readers can view the 3D object at once from different perspectives, enabling groups to share information in a virtual reality "world."

In 2003, the Sony Corporation patented a method of transmitting sensory data directly into the human brain. Sony's technology sends pulses of ultrasound through the cranium to modify brain waves, enabling the user to enjoy "artificial" sensory experiences including taste, sound, colors, and even moving images. A similar technology was used by the Riddler to read and control people's minds in the film *Batman Forever* (1995).

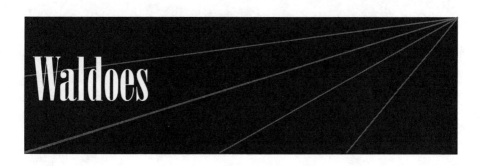

Waldoes

A "WALDO" IS A MECHANICAL HAND manipulated by a human operator, usually for purposes of handling dangerous materials in a laboratory. Perhaps the most famous fictional wearer of waldoes is Spider-Man's arch enemy, Dr. Otto Octavius, who controls four mechanical arms, fused to his body, with a waldo at the end of each.

Robert Heinlein invented the waldo in "Waldo" (1942). In the story, a scientific genius who is afflicted with a muscular disease, myasthenia gravis, puts himself into orbit; under zero gravity, he is weightless and therefore not physically limited by his muscular degeneration. The character performs experiments using remote-controlled mechanical hands, which are nicknamed "waldoes" after their inventor.

Global Shortage of Water

EARTH CAN BE CONSIDERED to be favored among planets in general, due to its abundance of water—a rare planetary occurrence that makes life possible. But could we ever run out of potable water? The possibility of global drought is more likely than you might imagine.

Every fifteen years, the world adds another billion people. The global population boom has caused demand for water to double over the last thirty years, according to *Utility Forecaster,* a newsletter that covers water companies and other utilities. Yet the amount of fresh water has remained pretty much constant.

The awful truth is that only one-third of the world's population has access to reliable sources of fresh drinking water. By 2025, the world will need up to three times the amount of clean, drinkable water that is currently available. According to the National Center for Atmospheric Research, the percentage of Earth's land area suffering drought has more than doubled in the past thirty years.

In China, home to 1.3 billion people, farmers pump thirty cubic kilometers more water to the surface each year than is replaced by rain. As water tables fall lower and lower throughout Asia, Vietnam has quadrupled the number of tube wells, from 250,000 to 1 million, over the past decade. An article in *NewScientist* (August 28, 2004) warns: "the world is on the verge of a water crisis as people fight over ever dwindling supplies."

The most important science fiction novel in which water—or more specifically, a shortage of water—plays a significant role is Frank Herbert's *Dune* (1965). Dune is a desert planet where water is so scarce, it is among the most precious of resources. On Dune, rich people don't show off by driving a Mercedes or BMW. The most conspicuous sign of wealth is, during a dinner party, to dump an entire pitcher of water on the ground.

Inhabitants of the planets wear "stillsuits," body-enclosing garments made of a micro-sandwich fiber. The microfibers help dissipate bodily heat and filter bodily wastes. Sweat and urine are filtered by the suit and made available as drinking water stored in catchpockets and sipped through a tube.

By volume, ninety-seven percent of the water on Earth is salt water; only three percent is fresh water. But most of that fresh water is frozen in polar ice caps and glaciers. Less than one percent of all the water on Earth is found in lakes, rivers, and under the surface.

In Kurt Vonnegut's 1963 novel *Cat's Cradle*, all the water on Earth becomes undrinkable when it permanently freezes into "ice nine," a form of ice that remains frozen at room temperature:

> There are several ways," Dr. Breed said to me, "in which certain liquids can crystallize—can freeze—several ways in which their atoms can stack and lock in an orderly, rigid way.
>
> "So it is with atoms in crystals, too; and two different crystals of the same substance can have quite different physical properties."
>
> He told me about a factory that had been growing big crystals of ethylene diamine tartrate. The crystals were useful in certain manufacturing operations, he said. But one day the factory discovered that the crystals it was growing no longer had the properties desired. The atoms had begun to stack and lock—to freeze—in a different fashion. The liquid that was crystallizing hadn't changed, but the crystals it was forming were, as far as industrial applications went, pure junk.
>
> How this had come about was a mystery. The theoretical villain, however, was what Dr. Breed called "a seed." He meant by that in tiny grains of the undesired crystal pattern. The seed, which had come from God-only-knows-where, taught the atoms the novel way in which to stack and lock, to crystallize, to freeze.

In J. G. Ballard's 1965 novel *The Drought*, radioactive waste from years of industrial dumping forms a protective crust over the oceans, interfering with water evaporation and weather patterns—it stops raining, and lakes and reservoirs are drying up, causing a global water shortage:

> At this point, attention switched to the ultimate source of rainfall—the ocean surface. It needed only the briefest scientific examination to show that here were the origins of the drought.
>
> Covering the off-shore waters of the world's oceans, to a distance of about a thousand miles from the coast, was a thin, but resilient

mono-molecular film formed from a complex of saturated long-chain polymers, generated within the sea from the vast quantities of industrial wastes discharged into the ocean basins during the previous fifty years.

This tough, oxygen-permeable membrane lay on the air-water interface and prevented almost all evaporation of surface water into the air space above. Although the structure of these polymers was quickly identified, no means was found of removing them.

The saturated linkages produced in the perfect organic bath of the sea were completely non-reactive, and formed an intact seal broken only when the water was violently disturbed. Fleets of trawlers and naval craft equipped with rotating flails began to ply up and down the Atlantic and Pacific coasts of North America, and along the seaboards of Western Europe, but without any long-term effects. Likewise, the removal of the entire surface water provided only a temporary respite—the film quickly replaced itself by lateral extension from the surrounding surface, recharged by precipitation from the reservoir below.

The mechanism of formation of these polymers remained obscure, but millions of tons of highly reactive industrial waste—unwanted petroleum fractions, contaminated catalysts and solvents—were still being vented into the sea, where they mingled with the wastes of atomic power stations and sewage schemes. Out of this brew, the sea had constructed a skin no thicker than a few atoms, but sufficiently strong to devastate the lands it once irrigated.

All life on Earth is dependent on water for survival. The average person needs slightly more than half a gallon of water a day. We get half of this by drinking water and other beverages; the rest comes from the moisture in foods we eat. What happens when we don't have water? A plum picked from a tree and left exposed to the Sun or wind becomes a prune. The dehydration of the plum produces the shriveled and wrinkled skin that is typical of a drying fruit. Loss of water causes the external and internal structures of living things to change, whether it's a prune, a raisin, a leaf—or you.

The largest supply of water on Earth is the oceans. They contain salt water, which is undrinkable. But what about converting that salt water to fresh water for drinking and bathing?

In his 1957 story, "The Man Who Ploughed the Sea," Arthur C. Clarke describes a "molecular sieve" that removes gold, uranium, and other minerals from seawater. Its main purpose is to harvest these minerals, but if all the minerals are removed, what's left is fresh water:

> They have made what we call a "molecular sieve." In its way, the thing *is* a sieve, and we can set it to select anything we like. It depends on very advanced wave-mechanical theories for its operation, but what it actually does is absurdly simple. We can choose any component of sea-water we like and get the sieve to take it out. With several units, working in series, we can take out one element after another. The efficiency's quite high, and the power consumption negligible.

A technology of this sort already exists and is in place: more than 12,500 commercial-scale desalination units, with an average production rate of 22.8 million cubic meters of water per day, are currently operating worldwide.

Salt can be removed from seawater in a number of different ways. One of the simplest, reverse osmosis, simply separates the salt from the water by filtering it through a fine membrane. Another widely used desalination technology is multistage flash desalination (MSF). The salt water is heated to a temperature at which it evaporates. The water forms condensate, leaving the salt behind. When the water is cooled, it is recovered as pure potable water.

While operational, desalination units are not yet an optimal solution to the challenge of producing more fresh water. The water they produce is not as pure as it could be, and the operational cost is rather high. Research to improve performance and lower cost is currently underway.

Weather Control

ALONG WITH EARTHQUAKES, tidal waves, fires, and volcanic eruptions, severe weather is one of the most devastating forces on Earth. No wonder we seek to control it, whether with technology or stories.

People tend to anthropomorphize weather; winter is symbolized by the image of "Old Man Winter," a slightly malevolent cloud with a human face whose breath causes blizzards. Native American "rain men" believed that by dancing and praying, they could bring rain in times of drought to save dying crops.

Humanity has long dreamed of controlling the weather. Samuel Johnson mentions weather control in *Rassek* (1759). Jane Loudon's *The Mummy* (1827) imagines weather control becoming a reality in a couple of hundred years. Francis Beeding's *The One Sane Man* (1934), John Boland's *White August* (1955), Ivan Yefremov's *Andromeda* (1959), Norman Spinrad's "The Lost Continent" (1970), Peter Dickinson's *The Weathermonger* (1958), Rick Raphael's "The Thirst Quenchers" (1963), and Ted Thomas' "The Weatherman" (1962) all deal with weather control.

In Ben Bova's *The Weathermakers* (1967) project THUNDER—Threatening Hurricane Neutralization Destruction and Recording—prevents hurricanes by warming up the air flowing *into* a storm and simultaneously cooling down the air at its center until temperatures throughout the air mass are uniform.

In the X-Men movies and comic books, Storm is a mutant whose power is the control over weather. And in the 1998 *Avengers* movie based on the 1960s television series, Sean Connery plays Sir August de Wynter, a villain who plans to rule the world with his weather control machine.

Scientists and meteorologists are constantly seeking new ways to monitor, predict, and even control the weather. Both the United States and Russia have reportedly conducted weather control experiments using extremely low frequency (ELF) electromagnetic waves. The theory is that the ELF waves will form major high-pressure systems that change the normal high-altitude jet stream patterns. Jet streams would be forced north by the ELF transmissions, blocking the normal flow patterns of incoming weather systems.

Of course, using science to control the weather is nothing new. Seeding clouds to make it rain began as early as 1946. Cloud droplets form around microscopic particles, including dust, soil, salt, and ice crystals. The idea is to place more particles into the cloud by dropping them from an airplane, so that droplets can form more rapidly, causing rain. A single cloud contains billions of tiny water droplets. The droplets are so small that it may take a million of them to form a single rain drop. The larger the concentration and size of droplets, the more readily it will rain. In warm weather, hygroscopic particles such as salt are used to seed clouds. A hygroscopic particle is a material that attracts and absorbs water. In cold weather, silver iodide or dry ice is used as the seeding material to increase rainfall.

According to Michel Chossudovsky of the University of Ottawa, the U.S. is developing weather control technology through a project called High-Frequency Actival Aural Research Program (HAARP). He writes, "Recent scientific evidence suggests that HAARP is fully operational and has the ability to potentially trigger floods, droughts, hurricanes, and earthquakes." HAARP allegedly consists of a system of powerful antennas capable of creating "controlled local modifications of the ionosphere"—the upper layer of the atmosphere. The antennas send out a powerful radio wave beam that heats the ionosphere. Since hot air rises, HAARP would lift areas of the ionosphere higher, changing local weather patterns.

But if there is a HAARP and it really works, why would the federal government not use the system to bring rain to areas of drought or even over raging forest and brush fires?

A fundamental principle of problem solving is that one must clearly understand and be able to state the problem before coming up with a solution. Weather is simply a pattern of atmospheric conditions: temperature, humidity, and wind velocity and direction. To understand problematic weather, such as hurricanes, scientists first create computer simulations of the hurricanes. Then, using computer modeling, they can determine the effect that changing some of the variables—for instance, increasing or decreasing the humidity—might have on the weather.

Computer simulations of hurricanes indicate that changes in precipitation, evaporation, and air temperature could change the path of the storm or slow the winds, diminishing the intensity. How can those changes be physically accomplished during an actual hurricane?

1. Cloud seeding with silver iodide could cause it to rain, removing the moisture hurricanes need to increase in intensity.
2. Another way to rob hurricanes of their "fuel" (moisture) would be to block evaporation of water over the path of the storm. For a hurricane forming over the ocean, biodegradable materials could be released, forming a membrane over the sea surface.
3. Satellites could beam microwaves into the storm, causing water molecules to vibrate and heat the surrounding air. The increased temperature could shift the path of the storm or lessen its intensity.

Wormholes (see also "Black Holes")

IN HIS BEST-SELLING BOOK *A Brief History of Time* (1988), Stephen Hawking points out that scientists, not science fiction writers, thought of the idea of wormholes first:

> The idea of wormholes between different regions of space-time was not an invention of science fiction writers but came from a very respectable source.
>
> In 1935, Einstein and Nathan Rosen wrote a paper in which they showed that general relativity allowed what they called "bridges," but Einstein-Rosen bridges didn't last long enough for a spaceship to get through: the ship would run into a singularity as the wormhole pinched off.
>
> However, it has been suggested that it might be possible for an advanced civilization to keep a wormhole open. To do this, or to wrap space-time in any other way so as to permit time travel, one can show that one needs a region of space-time with negative curvature, like the surface of a saddle.
>
> Ordinary matter, which has a positive energy density, gives space-time a positive curvature, like the surface of a sphere. So what one needs, in order to wrap space-time in a way that will allow travel into the past, is matter with negative energy density.

Einstein's field equations predicted that wormholes could be created by singularities, although their lifetime would be brief. The problem would be in enlarging them from their quantum size and then keeping them open.

The wormhole, as described by Hawking, could serve as a bridge or connection—possibly between a black hole and a white hole. A white hole is the opposite of a black hole: just as a black hole can only suck things in, a white hole can only spit things out.

There is nothing in Einstein's Theory of General Relativity that prohibits the existence of such white holes. These white holes would form like a collapsar in reverse, starting with a point of infinite density and ending with matter erupting outward into the universe. Thus, matter devoured by a black hole may be expelled by a white hole somewhere else in the universe. This smacks of being a sort of "space warp," where you go from one point to another instantaneously without having to travel between them.

Another model of white holes pictures matter from our universe passing through a black hole, and erupting through a white hole into an alternate universe. In this picture, the black hole/white hole pair forms a sort of "gateway" to another universe. Keep in mind, however, that the notion of an alternate universe is a mathematical model only. It can never be proven that it exists, since someone or something traveling through a black hole into such a universe would never be seen in our own world again.

In *Islands of Space* (1931), John Campbell used a "space warp" as a shortcut to get from one area of space to another (like the dots on a folded piece of paper). In his *The Mightiest Machine* (1934), Campbell referred to this interstellar shortcut as "hyperspace."

Isaac Asimov used "space jumps," a version of hyperspace, in many of his space novels. Other science fiction using similar devices include Robert Heinlein's *Sharman Jones* (1953), Murray Leinster's *The Other Side of Nowhere* (1964), and David Sandal's *Neverness* (1988).

In 1977, Adrian Berry suggested in his nonfiction book *The Iron Sun* that taking shortcuts through wormholes could save spaceships time during interplanetary travel. In the 1990 novel *The Ring of Charon* by Roger MacBride Allen, the entire planet Earth is captured and dragged through a wormhole.

In the 1991 novel *The Singers of Time*, by Frederik Pohl and Jack Williamson, astronauts use wormholes to travel between different universes. And in the TV series *Babylon 5* (1994), giant space stations could create and sustain wormholes to be used as portals to hyperspace.

X-Ray Vision

THE CLASSIC SCIENCE FICTION MOVIE of X-ray vision is *X: The Man with the X-ray Eyes* (1963). Ray Milland plays a scientist who invents eye drops to extend the range of his vision beyond the spectrum of visible light. He initially gains X-ray vision to amuse himself by seeing through people's clothes. But then he gets strange visions of other dimensions and other worlds, and the inability to turn off his extraordinary vision tortures him; the film ends with Milland ripping out his own eyes with his hands.

What, exactly, is an X-ray? Isaac Asimov, the master of explanation, provides this overview in his book *Words from Science* (1959):

> In 1895, the German physicist Wilhelm C. Roentgen noticed that a certain chemically coated paper in his laboratory glowed whenever his cathode-ray tube was in operation, even when there was cardboard between tube and paper. His cathode-ray tube would fog photographic plates, too, even when the plates were protected by their wrappings.
>
> He decided that some kind of radiation was formed in the cathode-ray tube that could pass right through glass, cardboard, and paper. He had no notion as to what this radiation might be, and he called them X-rays. This name suited the mystery surrounding the cathode rays, since X is the letter usually used by the mathematician to signify the unknown.
>
> Nowadays we know that the cathode rays are really streams of electrons, and we know that these X-rays are quite similar to ordinary light except that they are much more energetic.

Modern medical X-ray machines work in the same way, by fogging photographic plates. X-rays are high-frequency electromagnetic waves produced by a stream of electrons. As the X-rays pass through your body on the way to the plates, they are partially absorbed by organs and bones. Different body tissues absorb the X-rays in differing amounts. The pattern of "fogging" on the photographic plate is determined by the amount of X-rays that get through to it. Bones, which absorb X-rays the most, appear white on X-rays. Fat and other soft tissue absorbs less, and is seen as shades of gray. Air in your lungs hardly absorbs any X-rays, which is why lungs appear black on an X-ray.

Could a human be given the power of X-ray vision either through genetic engineering or optical implants? Doubtful: repeat exposure to the radiation from X-rays increases risk of cancer, which is why the dental technician leaves the room when he or she X-rays you; patients wear a lead apron during an X-ray for the same reason.

Superman had X-ray vision, but cancer was not a concern, because he was invulnerable to most forms of radiation, kryptonite being the major exception. And his X-rays, like any other, could not see through lead.

While you may never have Superman's X-ray vision, medical researchers are working on what may be the next best thing: a handheld "sonic flashlight" that could give doctors the equivalent of X-ray vision—without harmful X-rays. The sonic flashlight incorporates the same ultrasound technology used by gynecologists and obstetricians to show an image of a developing fetus inside a pregnant woman. Using a process called "tomographic imaging," a flat-paneled monitor with a half-silvered mirror allows the doctor to view the image at the same time he is pointing the sonic flashlight at the patient.

Appendix: The Runners-Up

A S I NOTED IN THE INTRODUCTION, for a science fiction concept
to make our list of "great science ideas that originated in science
fiction," the notion had to fit one of the following scenarios:

1. The idea clearly originated in science fiction, then was later discovered or achieved by scientists (e.g., putting a man on the Moon).
2. The idea originated in science, but was brought to the public's attention by science fiction writers (e.g., black holes).
3. The idea originated in science fiction, and we are getting increasingly close to making it a reality (e.g., flying cars, jet packs).
4. The idea originated in science fiction. It has not yet been realized in actuality by science. But it is theoretically possible according to current scientific findings (e.g., time travel, teleportation).

This, of course, forced us to leave off the list some absolutely fascinating science fiction ideas. In this appendix are a few of our favorite sf notions—major science fictional themes that are important and fascinating, but couldn't qualify for our list.

Faster-than-Light Travel

A LIGHT YEAR IS THE DISTANCE you can travel at the speed of light in one year. It is equal to 5.878 trillion miles. Einstein's theory states that nothing can travel faster than the speed of light. This limits a single human astronaut to destinations reachable within his normal life span—and practically, to distances much closer than that (no one wants to spend the bulk of his life traveling to reach a destination).

For intergalactic travel to be practical, we need to find a way to travel faster than the speed of light—a notion popularized by the warp factor of *Star Trek* and countless other science fiction TV shows, films, and stories. To calculate a starship's speed from its warp factor, you cube the warp factor number and multiply by the speed of light. So a ship traveling at warp factor 3 would be cruising at 3 x 3 x 3 = 27 times the speed of light. Was Einstein correct? Is the speed of light as we know it, 186,000 miles per second, the maximum velocity attainable?

Science fiction writers have devised a number of different ways to get around the light speed limit—some plausible, others fantastic. A favorite is to travel through a wormhole or other portal providing a shortcut between two points, circumventing normal space-time as we know it.

In other stories, matter can't exceed Einstein's speed limit, but energy can. Philip José Farmer's 1953 short story "Mother" refers to the "ultrarad," a communications device that sends faster-than-light waves through something called the "no-ether." Other faster-than-light communications devices in sf include James Blish's Dirac transmitter and Ursula K. Le Guin's "ansible," the operation of which is described in her novel *The Left Hand of Darkness* (1969):

> It doesn't involve radio waves, or any form of energy. The principle it works, on the constant of simultaneity, is analogous in some ways to gravity. What it does is produce a message at any two points simultaneously. Anywhere. One point has to be fixed, on a planet of a certain mass, but the other end is portable. That's this end. I've set the coor-

dinates for the Prime World, Hain. A NAFAL ship takes 67 years to go between Gethen and Hain, but if I write a message on that keyboard it will be received on Hain at the same moment as I write it.

E. E. "Doc" Smith got around the limiting speed of light for space travel by inventing what he called "inertia less travel." The idea is that if you deprive an object of inertia, theoretically it could be speeded up infinitely, surpassing light speed.

Some physicists now believe that the speed of light and other fixed numbers in the universe may in fact not be constant. If that's true, 186,000 miles per second may not be the permanent universal speed limit.

In gas clouds some twelve billion light years away from Earth, the "fine structure constant" seems to be slightly different than the fine structure constant of material in our own solar system. The fine structure constant is a measure of the electromagnetic forces that hold atoms together. Like c, the speed of light, the fine structure number is supposed to be constant throughout the universe. Therefore, if this fundamental constant can vary over time, there's no reason why the speed of light can't vary as well. In Vernor Vinge's *A Fire Upon the Deep* (1992), the speed of light varies depending on the region of space.

Also, a correct interpretation of Einstein's Theory of Relativity is not that you can't go faster than light, but that you can't break the light-speed barrier. So a spaceship can never be accelerated to a speed faster than light. It can come close to the light-speed limit, but never break it. However, there's nothing in Einstein's theory that forbids a particle of matter to travel faster than the speed of light at the instance of its creation. Such particles are called tachyons. Because of the restriction against crossing the light-speed barrier, a tachyon can never slow down and stop moving; it can decelerate to nearly the speed of light, but never less than that. Gregory Benford uses tachyons in *Timescape* (1980).

Another particle that may fulfill the promise of faster-than-light travel is the neutrino, a particle with zero mass. Tests to measure the speed of neutrinos, while inconclusive, indicate that if some neutrinos do indeed move faster than light, it is only by a small amount—a fraction of a mile an hour.

Even if tachyons do exist, a ship made out of ordinary matter (non-tachyons) cannot move faster than light. According to the theory of relativity, when an objective moves at a high velocity, its mass and length increase and time slows in relation to a stationary reference point. The mass of the object in motion can be calculated by using the following formula:

$$Mm = \frac{Mr}{(1-v^2/c^2)}$$

Where:

Mm = the mass of the object in motion
Mr = the mass of the stationery reference point object.
v   = velocity
c   = the speed of light.

As velocity gets closer to the speed of light, the denominator of the above equation, $1-v^2/c^2$, approaches zero, and the mass of the moving object approaches infinity. Since a single object in a multi-object universe cannot have one hundred percent of the mass in the universe, no object can attain light speed.

The bottom line is that, right now, faster-than-light travel looks as if it will have to forever remain merely a science fiction device. This speed limitation, in turn, severely limits the distances which manned space explorations can reach.

Force Fields

AS PORTRAYED IN SCIENCE FICTION, a "force field" is a wall of invisible force capable of deflecting both solid objects and energy. Isaac Asimov explained the concept more than sixty years ago in his 1941 short story "Not Final":

> All matter is composed of atoms. Atoms are held together by interatomic forces. Take away atoms. Leave interatomic forces behind. That's a force field.

The force field was the Robinson family's key means of defense in the 1960s television series *Lost in Space*. When threatened by aliens, the Robinson family would set up a force field projector in front of the spaceship, turn it on, and repel lasers, boulders, and other weapons and objects hurled at their ship.

Another famous science fiction force field is the "deflector" from *Star Trek*. As Lawrence M. Krauss explains in his book *The Physics of Star Trek* (1996):

> Warping space has other advantages as well. Clearly, if space-time becomes strongly curved in front of the *Enterprise*, then any light ray— or phaser beam, for that matter—will be deflected away from the ship. This is doubtless the principle behind deflector shields.
>
> Indeed, we are told that the deflector shields operate by "coherent graviton emission." Since gravitons are by definition particles that transmit the force of gravity, then "coherent graviton emission" is nothing other than the creation of a coherent gravitational field [which] is, in modern parlance, precisely what curves space. So once again the *Star Trek* writers have at least settled upon the right language.

While the Robinson's force field had to be set up on a stand by two men, and *Star Trek*'s deflectors were positioned on the front of a starship in a parabolic dish, Frank Herbert, in his classic novel *Dune* (1965), envisions a "personal" force field one can wear on the belt:

> Shield Defense: the protective field produced by a Holtzman generator. The field derives from Phase One of the suspensor-nullification effect. A shield will permit entry only to objects moving at slow speeds (depending on setting, this speed ranges from six to nine centimeters per second).
>
> Paul snapped the force button at his waist, felt the crinkled-skin tingling of the defensive field at his forehead and down his back, heard external sounds take on characteristic shield-filtered flatness. In shield fighting, one moves fast on defense, slow on attack. The shield turns the fast blow, admits the slow.
>
> Paul felt the field cracking as shield edges touched and repelled each other, sensed the electric tingling of the contact along the skin. The air within their shield bubbles grew stale from the demands on it that the slow interchange along barrier edges could not replenish. With each new shield contact, the smell of ozone grew stronger.

A similar personal force field is used in Charles Harness' "Flight into Yesterday" (1949), also published as *The Paradox Men* (1953).

The first mention of a force field may have been in E. E. "Doc" Smith's Skylark and Lensmen novels published in the 1930s and 1940s. Their use in *Star Trek* is quite similar to Smith's, suggesting that TV science fiction is about fifty years behind written sf. Other sf works involving force fields include Robert Sheckley's "Early Model" (1956) and Poul Anderson's *Shield* (1963).

While scientists have created many of the technologies envisioned by imaginative science fiction writers, force fields aren't one of them. The problem is that there is no known force capable of repelling all objects and energies. Gravity attracts, not repels, and no antigravity ray or generator has ever been developed. A magnetic force field might block metal weapons but would be useless against plastic.

As far as I can see, no one has developed even a prototype force field, and no one is even working on such a device. So it appears that the force fields predicted in science fiction won't become science fact in the foreseeable future.

Miniaturization and Shrinking Rays

A LTHOUGH SCIENCE FICTION has often been described as a literature of big ideas, one of the science fiction writer's favorite ideas is actually very small: miniaturization.

Miniaturization, as it is used in science fiction, refers to a shrinking machine or other technology, such as a chemical or radiation treatment, that causes an object to become significantly smaller.

In the 1940 science fiction film *Dr. Cyclops*, a mad scientist with a laboratory deep in the Peruvian jungle invents an atomic shrinking ray. He uses it to shrink several people to six inches in height and makes them his prisoners.

In *The Incredible Shrinking Man*, a 1957 movie based on a short story by Richard Matheson, Scott Carey is exposed to a strange mist and begins to shrink. Carey becomes so small that he lives in a dollhouse and has to battle for his life against common household and backyard insects. Eventually he shrinks to atomic size and smaller and discovers that atoms are self-contained universes where life exists on a smaller scale. When he lands on a planet (electron) in the subatomic universe, he is a giant. Eventually he becomes tiny even on this planet, as he continues to shrink.

In his 1966 novel *Fantastic Voyage*, Isaac Asimov gives a plausible-sounding explanation of a miniaturization process:

Miniaturization is quite possible. Have you ever seen a photograph enlarged or reduced to microfilm size?

Without theory, then, I tell you that the same process can be used on three-dimensional objects; even on a man. We are miniaturized, not as literal objects, but as images; as three-dimensional images manipulated from outside the universe of space time.

What physicists discovered ten years ago was the utilization of hyper-space; a space, that is, of more than the three ordinary spatial dimensions. The concept is beyond grasp; but the funny part is that it can be done. Objects can be miniaturized. We reduce the size of the atoms, too; we reduce everything; and the mass decreases automatically. When we wish, we restore the size.

In principle we can reduce a man to the size of a bacterium, of a virus, of an atom. There is no theoretical limit to the amount of miniaturization. We can shrink an army with all its men and equipment to a size that will fit in a match-box. Ideally, we could then ship that match-box where it is needed and put the army into business after restoring it to full size.

With nanotechnology, we have made miniaturization—in the sense of being able to build things smaller and smaller—come true. But no one has even come close to developing a miniaturization process or shrinking ray, as inventor Wayne Szalinski does in the movie *Honey, I Shrunk the Kids* (1989):

"The machine uses two beams, a laser-tracking beam and the actual shrinking beam. The tracking beam focuses the second beam on the target. The reason things have been blowing up instead of shrinking is that when the beams collide, they generate too much heat. Yesterday, this ball must have blocked the tracking beam, and the shrinking beam worked all on its own. Now, all I have to do is shut off the one beam"—He threw the switch into the "On" position. The machine came to life with a hum, the beams jumping into space.

In the DC Comics comic book, *The Atom*, Professor Ray Palmer finds a fragment of a white dwarf star that crashed to Earth as a meteor. When ultraviolet light is reflected off the white dwarf fragment, it creates a shrinking ray (presumably because it is from a "dwarf" star). Palmer creates a special lens that controls the ray, giving him the ability to shrink to atomic size and return to normal size at will.

Monsters

ONSTERS ARE THE THEME at which horror and science fiction intersect. The wolfman is a monster, and he's definitely part of the horror genre. But what about Frankenstein's monster? We think of him as a classic horror movie monster, yet his creation—bringing life to an inanimate body using electricity—is pure science fiction.

The monsters in science fiction are generally of two types: nonhuman creatures and men transformed into monsters.

In the nonhuman creature sector are such sf favorites as Godzilla, Rodan, and the Creature from the Black Lagoon. Godzilla is a giant lizard who is the size of a tall building and shoots atomic fire out of his mouth. Rodan is a giant pterodon. The Creature from the Black Lagoon is some sort of missing link between fish and man.

Giant bugs are central to the plot line in numerous science fiction films, and you can guess what kind of bug starred in each just from the titles: *Earth vs. the Spider* (1958), *The Black Scorpion* (1957), *The Deadly Mantis* (1957), and *Tarantula* (1955), which features Leo G. Carroll, who later played Mr. Waverly in *The Man from U.N.C.L.E.* television series as a scientist who transforms into a monster after he injects himself with a serum.

Not all insect monster movies reveal the nature of the beast in the title. In *Beginning of the End* (1957), Peter Graves, best known as Mr. Phelps from the *Mission Impossible* TV series, has to help Chicago get rid of giant grasshoppers. In *Monsters from Green Hell* (1957), the threat is wasps that have grown to gigantic size from exposure to solar rays.

Monsters from the sea are also a recurring plot line in sf films. In *It Came from Beneath the Sea* (1955), Army soldiers battle a giant octopus as it destroys the Golden Gate Bridge. In *The Beast from 20,000 Fathoms* (1953), nuclear tests wake a Godzilla-like dinosaur from its hibernation in a glacier at the North Pole. The beast wreaks havoc making its way to its ancestral breeding ground, which happens to be the portion of the Hudson River in which Manhattan is located. Godzilla himself comes to Manhattan to breed by way of the Hudson River in the 1998 *Godzilla* remake starring Matthew Broderick.

Perhaps the most famous nonhuman monster would have to be the Loch Ness monster. Rumors of the monster began when, in 1934, a London physician allegedly photographed a dinosaur-like beast with an elongated neck emerging from the waters of Loch Ness, a long, deep lake near Inverness, Scotland.

Since then, many sightings have been reported. Some speculate that the Loch Ness monster is a prehistoric animal (see "Prehistoric Creature") that survived extinction when a group of them swam into Loch Ness from the ocean and then became trapped and unable to get out. But recent submarine explorations of the lake have found no clear signs of an animal that size, and worse, they show there is not enough vegetation to support a family of such creatures. One investigator suggested that the Loch Ness monster could be a Baltic sturgeon, a fish that can grow up to nine feet long and weigh as much as 450 pounds.

Other famous monsters include the Jersey Devil, a hybrid creature with wings and claws, believed to live in New Jersey's Pine Barrens. Legend has it that the Jersey Devil was the thirteenth child of a thirteenth child, and the combination of bad luck produced a monstrosity.

Then there is Bigfoot, also known as the Yeti, and the Abominable Snowman. There have been many reported sightings and blurry photos taken, but none of these beasts, thought to be a sort of "missing link," has been captured, and there is no hard scientific evidence that they exist.

The classic sf monster story of all time, written in 1938 by John W. Campbell (under his pen name Don Stuart), is probably "Who Goes There?" A being is found frozen in polar ice. When thawed out, it revives and terrorizes the humans who found it. With the ability to control its individual cells at will, the creature can copy another organism's cell structure, becoming a duplicate of that animal or person.

As for men who become monsters, these are typically scientists who are experimenting with animals and inject themselves with serums that unexpectedly transform them into half-men, half-beasts.

In the 1896 H. G. Wells novel *The Island of Dr. Moreau*, a mad scientist performs surgery on jungle animals to make them more intelligent and human, creating a village of deformed man-beasts:

> These creatures you have seen are animals carved and wrought into new shapes. To that—to the study of the plasticity of living forms—

my life has been devoted. I have studied for years, gaining in knowledge as I go.

I see you look horrified, yet I am telling you nothing new. It all lay in the surface of practical anatomy years ago, but no one had the temerity to touch it. It's not simply the outward form of an animal I can change. The physiology, the chemical rhythm of the creature may also be made to undergo an enduring modification, of which vaccination and other methods of inoculation with living or dead matter are examples that will, no doubt, be familiar cases.

Less so, and probably far more extensive, were the operations of those mediæval practitioners who made dwarfs and beggar cripples and show-monsters; some vestiges of whose art still remain in the preliminary manipulation of the young mountebank of contortionist. Victor Hugo gives an account of them in *L'Homme qui Rit.*

But perhaps my meaning grows plain now. You begin to see that it is a possible thing to transplant tissue from one part of an animal to another or from one animal to another, to alter its chemical reactions and methods of growth, to modify the articulations of its limbs, and indeed to change it in its most intimate structure.

And in his 1985 short story "Mengele," Lucius Shepard describes a village populated by the victims of surgical experiments performed by the Nazi scientist Dr. Joseph Mengele:

> ...oblate heads, strangely configured hands, great bruised-looking eyes that seemed patches of velvet woven into their skins rather than organs with humors and capillaries.

In the 1959 film *The Alligator People*, a scientist hopes to harness the regenerative power of lizards to grow new arms and legs on people missing a limb. Unfortunately, his formula turns them into alligator creatures. Dr. Curt Connors, a scientist in the *Spider-Man* comic books, performs a similar experiment and accidentally transforms himself into the Lizard, a half-man, half-reptile creature. And in the 1959 film *The Hideous Sun Demon*, a scientist transforms into a scale-covered beast when exposed to sunlight.

The most famous film of human turned into animal/human hybrid is *The Fly* (1958). Instead of a serum, a scientist invents a teleportation device, a precursor of *Star Trek's* transporter. In one trial run he performs on himself, a fly accidentally gets into the booth with him. When they emerge in the receiving transporter chamber, he has a fly's head, though he retains much of his human intelligence, and the fly has his head.

Realizing he has become a monster and that his mind is being taken over by the fly's bestial instincts, the scientist crushes himself to death in a hydraulic press. Later, a colleague discovers the fly with his human head trapped in a spider web. In a classic scene, the terrified fly/human hybrid shrieks, "Help me! Help me!" Horrified, the colleague smashes the monstrosity, the spider, and the web with a rock.

Sometimes the creation of a human monstrosity is caused not by a scientific experiment gone wrong, but by an alien life-form that possesses an astronaut or someone who happens to find a meteor in a field. In *The Creeping Unknown* (1955), an alien life-form turns an astronaut into a giant fungus-like blob that absorbs the life essence from other living things. In Stephen King's 1971 story "I Am the Doorway," an alien creature infecting a human being begins to erupt from his body as eyes on his hands.

Are monsters real? There are no werewolves, vampires, or other human monsters, but there certainly are human abnormalities which we unkindly call "freaks." Siamese twins, where two twins do not fully separate before being born, are the most common. Such Siamese twins may be attached at the head or torso, share internal organs such as a brain or liver, and also share limbs. Sometimes twins are born that are not even two distinct people. The larger, normal-sized twin has a normal body, but a smaller, undeveloped twin is attached to that body, often as a second pair of legs, or as in some cases, a small body protruding from the abdomen. Some cases of "Cyclops" babies—infants with a single eye in the middle of the head—have been reported, but these rarely live more than a few hours.

Sometimes "monsters" are created by disease or heredity. Grady Stiles, known in sideshows as "Lobster Boy," came from a family whose members were born with their fingers fused together so they looked like lobster claws. People with the condition ichthyosis, causing a dry scaly skin, have worked as "alligator men" in sideshows because of their lizard-like appearance.

There are no werewolves, but the condition of "lycanthropy"—believing that you are a wolf—is common enough that it appears in *Dorland's Illustrated Medical Dictionary*. And over the past few years, numerous feature stories have been written about groups of men and women who call themselves vampires and, although they are not supernatural beings, drink human blood.

Perpetual Motion Machines

A S THE NAME IMPLIES, a perpetual motion machine is a machine that moves perpetually—that is, it never stops moving. Because the movement in such a machine is a form of kinetic energy, the perpetual motion machine can be said to generate perpetual (and hence unlimited) energy. For instance, if a pendulum swings forever, you can connect the pendulum to a gear to drive an armature through a magnetic field, and keep it spinning perpetually to generate an endless supply of what would essentially be free electricity.

In his early stories, John W. Campbell described a technology called "molecular motion" that converted the thermal energy of any substance into linear motion. It's not quite perpetual motion, though, because when the substance runs out of thermal energy and reaches zero degrees, the linear motion stops.

To generate free energy, a perpetual motion machine must move, continually, without the aid of a power source. In 1618, Dr. Robert Fludd announced that he had invented such a perpetual motion machine. Fludd's machine was a waterwheel that did not need a running river or canal to drive it. Water was poured into the machine from a bucket or reservoir. The water turned the wheel, providing power to a pump. The pump in turn caused the water to flow back over the wheel, repeating the cycle. A shaft connected to the water wheel rotated as the wheel did, turning gears to grind grain. So why didn't it work? There is no such thing in the universe as free energy. Friction created by the wheel and pump turn some of the kinetic energy into heat and noise. What is left over is not enough to both grind the grain and keep the wheel going indefinitely.

Today some physicists are investigating what may in fact be a free source of energy. Called "zero point energy," the process converts high-frequency electromagnetic radiation to electrical energy. Zero point energy results from the laws of quantum mechanics. In classical physics, thermal energy, or heat, is generated by the movement of atoms or molecules in a substance. We call this "Brownian motion." When there is zero molecular movement, there is no heat, and you have reached the temperature of absolute zero.

But even when the molecules and atoms are at rest, there is a slight movement or vibration of the subatomic particles of which they are composed. No object can ever have precise values of velocity and position at the same time; hence nothing can be perfectly still (the law of quantum mechanics).

If there is still movement even at absolute zero—the "zero point"— then there is still thermal energy—"zero point energy." If we can find a way to convert this zero point electromagnetic energy into electrical energy, we will in essence have free energy from a perpetual motion machine: the perpetual motion of atoms at the quantum level, even at absolute zero.

Of course, if you succeed in building and patenting a perpetual motion machine, you will become rich beyond the dreams of avarice. But don't send a drawing of your invention to the U.S. Patent Office. Because they believe the idea of a perpetual motion machine is patently ridiculous, they'll only accept a working model for submission—no drawings allowed.

Star Wars

HE SCIENCE FICTION IDEA of wars between star systems began about the time Hubble demonstrated the existence of other galaxies and that the universe was expanding. E. E. "Doc" Smith wrote about such conflicts in *The Skylark of Space* (1928).

About the same time, Edmond Hamilton was publishing a series of stories in *Weird Tales* featuring the Interplanetary Patrol, such as "Crashing Suns" (1928). He became known as "World-Destroyer," "World-Wrecker," or "World-Saver" Hamilton.

John W. Campbell also took his place among the authors of space epics and wars between star systems with stories collected in *The Black Star Passes* (1953), *Islands of Space* (1931), *Invaders from the Infinite* (1932), and *The Mightiest Machine* (1934).

Jack Williamson is noted for his The Legion of Space trilogy (1934, 1936, 1939), with the ultimate (and simple) weapon, AKKA. Many fictional wars between star systems have happened since, including such catastrophic scenarios as George Zebrowski's and Charles Pellegrino's *The Killing Star* (1995) and Greg Bear's *The Forge of God* (1987) and *Anvil of Stars* (1992). And, of course, *Star Wars* features conflicts between star systems.

When Ronald Reagan was president, his Strategic Defense Initiative, known as "Star Wars," frequently made the news. But, unfortunately for star wars' place in this book, it had nothing to do with fighting aliens.

Superconductivity

WHEN I BEGAN RESEARCHING THIS BOOK, I was certain that superconductivity would make it on our list of "the greatest science ideas to originate in science fiction." So why is superconductivity relegated to this appendix as a "runner-up"?

For the simple reason that I could find only a few major references to superconductivity in early science fiction writings. Since the discovery of superconductivity by scientists, it has made its way into sf stories, but more as minor plot device than a major theme. For instance, in Roger Zelazny's "Go Starless in the Night" (1979), a terminally ill patient cryogenically frozen at the moment of death gains consciousness when his body temperature is lowered to near absolute zero, transforming his nervous system into a superconductor.

The closest link between superconductivity and science fiction that I could discover is science fiction writer James C. Glass. Like many sf authors, Glass is also a scientist, and as a professor of physics, his research specialty was superconductivity. An active science fiction writer, Glass' books include *Shanji* (1996) and *Matrix Dreams and Other Stories* (2004).

So...what is superconductivity? Well, a "conductor" is a material, like metal, which conducts electricity. An insulator, on the other hand, is a material through which electricity cannot flow. The higher the resistance of a material, the better insulation it makes. Rubber is an example. Materials with lower resistance, like copper, make good conductors. But even conductors have some resistance; they are not "zero resistance" materials.

As the name implies, "superconductivity" refers to a material which conducts electricity far more readily than a conventional conductor, which implies a very low resistance. Superconductivity is typically achieved only at extremely low temperatures. The lowest possible temperature is absolute zero, which is zero degrees on the Kelvin temperature scale and minus 454 degrees on the Fahrenheit scale. At absolute zero, molecules stop vibrating entirely, and the only motion that occurs is on a quantum scale. Because of the quantum motion, an effect of the Uncertainty Principle, the vibrations in matter can never be totally eliminated, and therefore you can't reach zero heat and absolute zero in temperature.

Scientists at the University of Maryland have successfully cooled a tiny piece of material, gold and silver nitride, to within sixty millikelvins, or six thousandths of a degree Kelvin, above absolute zero.

In 1911, the Dutch physicist Heike Kamerlingh Onnes discovered that mercury loses its resistance to an electrical current at very low temperatures. Further research showed that many other metals also became superconductive, with zero electrical resistance, at very low temperatures.

In 1950, John Bardeen found that isotopes, which are different forms of the same element, become superconductors at different temperatures. The reason: electrical conductivity is caused when quantums of vibrational energy disrupt the flow of electrons that we know as electricity. At extremely low temperatures, these quantum vibrations are reduced, and electrons are able to form pairs in which two electrons have opposite spin and momentum. When current is applied, the electron pairs move through the superconducting material without resistance.

The trouble with superconductivity is that it requires cryogenic temperatures that are difficult to produce and sustain. Superconductivity research today focuses on finding materials that are superconducting at higher temperatures, called *high-temperature superconductivity (HTS)*. Traditional low-temperature superconducting materials must be cooled with liquid helium, which is costly to produce. HTS materials can be adequately cooled with less costly liquid nitrogen.

Mercury-based cuprate can be superconducting at 130 degrees Kelvin, the highest recorded temperature at which an HTS material can maintain zero resistance. Barium-doped lanthanum copper oxide becomes superconductive at thirty-six degrees Kelvin. Magnesium diboride is superconductive at forty degrees Kelvin. Liquid nitrogen, by comparison, has a temperature of seventy-seven degrees Kelvin.

Many other HTS materials have been discovered. These include iron, single crystals of carbon-60, DNA, and crystalline organic materials.

Materials that are superconductive at high temperatures are being developed by American Superconductor, a publicly traded company. Their product, a high-temperature superconductive wire, saves electricity and is simple to install. Because electrical resistance is zero, power utilization through the wire is far more efficient than in conventional copper wire. Other applications of HST materials include sensors, MRI equipment, computer chips, and cell phone base stations.

Researchers at Brookhaven National Laboratory in Long Island, New York, have developed a new superconductor, sodium cobalt oxyhydrate, that can be easily manufactured in large quantities without generating toxic byproducts. This metal oxide has a complex structure in which thin layers of cobalt oxide are separated by layers of water molecules with sodium ions.

Superintelligent Animals

SUPERINTELLIGENT ANIMALS occur with considerable frequency in fantasy (dragons, cats, werewolves); not as often in science fiction. Olaf Stapledon's *Sirius* (1944) features a superintelligent dog, as does Harlan Ellison's story "A Boy and His Dog" (1969), in which the dog, named Blood, is also telepathic.

The apes in Edgar Rice Burroughs' Tarzan novels display intelligence greater than a real ape's. Sometimes superintelligent animals work with humans as part of a team, such as the superintelligent mutated Kodiak bears in Murray Leinster's "Exploration Team" (1956).

In Pierre Boulle's *Planet of the Apes* (1963), apes evolve until they are as intelligent as humans and eventually become our masters. The 2001 remake of *Planet of the Apes* makes it clearer than the original why such apes would rule us: even if our intelligence is on the same level, an ape is eight times stronger than an average human, making them able to easily dominate and defeat us physically.

In his novella *Flowers for Algernon* (1959), Daniel Keyes imagines increasing the intelligence of animals through surgery. A laboratory mouse, Algernon, has his intelligence enhanced through this procedure. Eventually the operation is performed on Charlie Gordon, turning the retarded man into a genius.

The farmyard animals in George Orwell's 1945 novel *Animal Farm* revolt against the human farmers, but soon the Pigs take control and the animals are no better off than they were before they revolted.

Other fictional superintelligent animals include the dogs and gorillas in Lester del Rey's "The Faithful" (1938), the dogs in Clifford Simak's *City* (1952), the dolphins in Gordon R. Dickson's "Dolphin's Way" (1964), the whales in Ian Watson's *The Jonah Kit* (1975), the cats in Cordwainer Smith's "The Game of Rat and Dragon" (1955), the dolphins in Robert Merle's *The Day of the Dolphin* (1969), and the dolphins and chimpanzees in David Brin's *Startide Rising* (1983), in which these animals are "uplifted" to sentience by humans. Even Matt Groening's

The Simpsons features an episode in which superintelligent dolphins revolt and take over Springfield.

So, have animals gotten smart enough to boss us around yet? And will they ever?

Both dolphins and apes have exhibited signs of high intelligence. A dolphin's brain weight averages three-and-one-half pounds, half a pound heavier than a man's. In an experiment, a dolphin raised in captivity was trained to come to the surface and make a sound when researchers spoke to him. They found that eighteen percent of the sounds the dolphin made duplicated sounds he heard from humans, whose language he was trying to imitate.

Recently, marine biologists have discovered that dolphins are capable of using tools. The dolphins break marine sponges off the sea floor and wear them over their snouts while foraging. Says one observer, "We believe that they use sponges as a kind of glove to protect their sensitive rostrums while they probe for prey."

Then there is Koko, the world-famous gorilla. She can communicate with humans using sign language, in which she knows more than a thousand words. Capable of love and other high-level emotions, Koko once adopted a kitten that found its way into her cage.

Bruce Lahn, a biomedical researcher at the University of Chicago, says he has found the gene responsible for humans developing their big brains and superior intelligence: the *abnormal spindle-like microcephaly associated (ASPM)* gene. The gene controls how many times cells in the cerebral cortex can divide, which in turn determines how many neurons are in the brain. Lahn plans to insert the ASPM gene in laboratory mice to see whether it affects their brain development and makes them smarter.

Utopian Societies

NE OF THE EARLIEST WRITERS to envision a Utopian society was Plato, a Greek philosopher who lived three hundred to four hundred years before the birth of Christ. His satirical utopia, known as "Plato's Republic," was a city-state with communal living among the ruling class.

The writers of the Bible also imagined what is the best-known of the Utopian societies, the Garden of Eden. Adam and Eve live in Eden in bliss until a serpent tricks them into disobeying God, who casts them out as punishment for eating forbidden fruit that gave them knowledge.

Thomas More wrote about his own version of utopia in his book *Utopia*, published in 1516. In the utopian society envisioned by More, all land is owned by the city-state and there is no private property. There is no money and no wealth. The city-state is ruled by a prince who is elected to rule for life.

More was the first writer to use the word "utopia." Other early utopias: Thomas Campanella's "The City of the Sun" (1623), Francis Bacon's "The New Atlantis" (1627), Samuel Butler's *Erewhon* (1872), Edward Bellamy's *Looking Backward: 2000–1887* (1888), and Eric Frank Russell's "...And Then There Were None" (1951).

H. G. Wells wrote several books about future utopian societies including *The Shape of Things to Come*, published in 1933. Also published in 1933 was James Hilton's classic novel *Lost Horizon*, in which survivors of a plane crash are rescued and taken to Shangri-La, a hidden utopian society in which people remain eternally youthful.

In *The Time Machine* (1895), Wells describes a future utopian society which turns out not to be as ideal as it seems—a common theme in science fiction novels about utopias:

> So far as I could see, all the world displayed the same exuberant richness as the Thames valley. From every hill I climbed I saw the same

abundance of splendid buildings, endlessly varied in material and style, the same clustering thickets of evergreens, the same blossom-laden trees and tree-ferns. Here and there water shone like silver, and beyond, the land rose into blue undulating hills, and so faded into the serenity of the sky.

I must confess that my satisfaction with my first theories of an automatic civilization and a decadent humanity did not long endure. Yet I could think of no other. Let me put my difficulties. The several big places I had explored were mere living places, great dining-halls and sleeping apartments. I could find no machinery, no appliances of any kind.

Yet these people were clothed in pleasant fabrics that must at times need renewal, and their sandals, though undecorated, were fairly complex specimens of metalwork. Somehow such things must be made. And the little people displayed no vestige of a creative tendency. There were no shops, no workshops, no sign of importations among them. They spent all their time in playing gently, in bathing in the river, in making love in a half-playful fashion, in eating fruit and sleeping. I could not see how things were kept going.

Unfortunately, Wells' narrator was correct in his misgivings. It turns out humankind had split into two races. The surface dwellers, or Eloi, live on the surface in a sunlit Utopia without a care in the world. The subterranean Morlocks provide for all their needs and do all the work, but not without adequate payment: the Morlock treat the Eloi as cattle, periodically coming to the surface to capture some Eloi, take them underground, and eat them.

After World War I, the antiutopia, negative response to utopian ideas—and the dystopian—became more common: Aldous Huxley's *Brave New World* (1932), George Orwell's *1984* (1949), and John Brunner's *Stand on Zanzibar* (1968). Frederik Pohl's and Cyril Kornbluth's *The Space Merchants* (1952) is a dystopia dominated by advertising agencies.

Many of the utopias (and some of the dystopias) are antiscience. They attribute the ills of society to humanity's growing dependence on machinery—as in E. M. Forster's "The Machine Stops" (1909)—and advocate a return to a simpler agrarian society.

There were many nineteenth-century attempts to set up utopias, such as Brook Farm (1844–1846), which enlisted many prominent transcendentalists. Amos Bronson Alcott (Louisa May's father) attempted a vegetarian utopian commune in 1843.

The Shakers set up some religious utopian communities, written about by Gerald Jonas—the sf reviewer for many years for the *New York Times*—in "The Shaker Revival" (1970). Other religious groups set up their own communities, like the Amana Society of Iowa, to keep them from being contaminated by corrupt society.

Introduction

Gunn, James, "Science, Science Fiction, and the Future," Welcoming Remarks at the 2004 Campbell Conference.

Rabkin, Eric, "Novel Inspiration," *BusinessWeek*, 10/11/04, p. 206.

Vergano, Dan and Watson, Traci, "Capsule Bearing Solar Dust Crashes," *USA Today*, 9/9/04, p. 1.

Alternate Energy

"Electricity from Wind," Power Scorecard, www.powerscorecard.org.

"HDR Geothermal Energy in Australia," www.geodynamics.com.

"Hydro-electricity," www.pge.edvcs.com.

"Introduction to Concentrating Solar Power," U.S. Department of Energy, www.nrel.gov/clean_energy.

"Introduction to Geothermal Electricity Production," U.S. Department of Energy, www.nrel.gov/clean_energy/geoelectricity.html.

"It Wasn't All Bad," *The Week*, 1/28/05, p. 4.

"New Wave-Pump Technology is Successfully Demonstrated," *Chemical Engineering Progress*, 2/04, p. 8.

"Power Sludge," *Scientific American*, 5/04, p. 38.

"Silicon Solar Cell," *Scientific American*, 6/04, p. 21.

"Wind Power Changes Weather," *NewScientist*, 11/19/04, p. 22.

"Wind Power's Imprint on the Environment," *BusinessWeek*, 11/22/04, p. 127.

www.gasification.org

Chang, Kenneth, "New Fusion Method Offers Hope of New Energy Source," *New York Times*, 4/8/03.

Doyle, Alister, "Moon Brings Novel Green Power to Arctic Homes," *Reuters*, 9/23/03.

Fridleifsdottir, Siv, "The Renewable Energy Century," Ministry for the Environment, Iceland.

Graham-Rowe, Duncan, "Hydro's Dirty Little Secret Revealed," *NewScientist*, 2/26/05, p. 8.

Graham-Rowe, Duncan, "Making the Best of Garbage Gas," *NewScientist,* 2/26/05, p. 25.

Halacy, D. S., *Solar Science Projects For a Cleaner Environment* (Scholastic Book Services, 1971).

Llanos, Miguel, "Poop Power," MSNBC, 7/16/04.

Mandelbaum, Robb, "Greenmark," *Discover,* 6/04, pp. 50–55.

Nelson, Jennifer, "Getting a High-Voltage Charge from the Sea," *NJ Business,* 9/22/03, p. 8.

Pearce, Fred, "Tear Off a Sheet of Solar Cells," *NewScientist,* 12/18/04, p. 23.

Whitehouse, David, "Fusion Power Within Reach," *BBC News,* 10/1/01.

Alternate Universes

Holt, Jim, "My So-Called Universe," *Slate,* 8/20/03.

Overbye, Dennis, "A New View of Our Universe: Only One of Many," *New York Times,* 10/29/02.

Tegmark, Max, "Parallel Universes," *Scientific American,* 5/03.

Whitehouse, David, "Before the Big Bang," *BBC News Online,* 4/10/01.

Androids

www.tvacres.com/robots_androids.htm

Clute, John and Nicholls, Peter, *The Encyclopedia of Science Fiction* (St. Martin's, 1995).

Trimble, Bjo, *Star Trek Concordance* (Citadel Press, 1995).

Antigravity

"LIGO," *NewScientist,* 8/28/04, p. 33.

Dvali, Georgi, "Out of the Darkness," *Scientific American,* 2/04, p. 70.

Sica, R.J., "A Short Primer on Gravity Waves," Department of Physics and Astronomy, The University of Western Ontario.

Wesson, Paul, "The Light Stuff," *NewScientist,* 11/20/04, p. 31.

Wright, Karen, "Black Holes Made Here," *Discover,* 6/04, pp. 62–63.

Antimatter

www.matter-antimatter.com

http://livefromcern.web.cern.ch/livefromcern/antimatter/

Collins, Gerald, "Making Cold Anti-Matter," *Scientific American,* 6/05, pp.55–63

Concise Science Dictionary (Oxford University Press, 1991).

Mann, George, *The Mammoth Encyclopedia of Science Fiction* (Carroll & Graf, 2001).

Artificial Intelligence

"Intel Ships Pentium 4 Processor Operating at 2.2 Billion Cycles per Second," www.intel.com/pressroom.

"Kasparov vs. Big Blue: The Rematch," *SIAM News*, 6/97.
"Robots Reason Scientifically," *Chemical & Engineering News*, 1/19/04, p. 62.
"Take One DNA Computer and Heat Gently," *NewScientist*, 6/11/04, p. 21.
www-formal.standford.edu/jmc/whatisai/node1.html
www.genetic-programming.org
Adler, Irving, *Thinking Machines* (New American Library, 1961).
Graham-Rowe, Duncan, "Glooper Computer," *NewScientist*, 3/26/05, p. 33.
Nolte, David, *Mind at Light Speed* (Free Press, 2001).
Koza, John; Keane, Martin; and Streeter, Matthew, "Evolving Inventions," *Scientific American*, 2/03, pp. 52–59.
Katz, John, "Can Androids Feel Pain?" Slashdot, http://slashdot.org/features.
McDougall, Paul, "Feds Take Two Routes to Supercomputer Power," *Information Week*, 8/2/04, p. 30.
Ricadela, Aaron, "Petaflop Imperative," *Information Week*, 6/21/04, p. 55.
Robinson, Sara, "Human or Computer?" *New York Times*, 12/10/02.
Rovin, Jeff, *Classic Science Fiction Films* (Carol Publishing, 1993).
Walsh, Nick, "Alter Our DNA or Robots Will Take Over," *The Observer*, 9/2/01.

Artificial Life

www.chem.duke.edu.
www.newscientist.com.
Holmes, Bob, "Alive," *NewScientist*, 2/12/05, pp. 29–33.
Crichton, Michael, *Prey* (HarperCollins, 2000).
Kroeker, Kirk and Vos Post, Jonathan, "Writing the Future: Computers in Science Fiction," *IEEE Computer*, 1/00, pp. 29–37.
Ball, Philip, "Life's Cycle," *Nature*, 7/21/00.
Gibbs, W. Wayt, "Synthetic Life," *Scientific American*, 5/04, pp. 75–81.
Gillis, Justin, "Scientists Plan to Create New Organism In Lab Dish," *The Record*, 11/22/02.
Ward, Mark, *Virtual Organisms* (St. Martin's, 1999).
Zimmer, Carl, "What Came Before DNA," *Discover*, 6/04, pp. 34–41.

Asteroids Colliding with the Earth

"A Vast Field of Craters," *The Week*, 11/26/04, p. 21.
"Close Calls," *Scientific American*, 5/04.
"Dual Asteroid Strike Hits Comet Theory," *NewScientist*, 7/31/04, p. 8.
"Nuclear Close Call," *Scientific American*, 1/03.
Britt, Robert, "Small Asteroid Zooms Past Earth," Space.com, 3/18/04.
Choi, Charles, "Permian Percussion," *Scientific American*, 7/04, p. 36.
Ferris, Timothy, "Killer Rocks from Outer Space," *Reader's Digest*, 10/02.
Jaroff, Leon, "Will a Killer Asteroid Hit the Earth?" Time.com, 4/3/00.
Long, Wei, "China Builds New Observatory to Detect Near-Earth Asteroids," space.com, 8/15/00.

McCall, William, "Meteor Study Finds Risk Overestimated," *The Record*, 11/21/02.

Steel, Duncan, *Target Earth* (Reader's Digest, 2000).

Whitehouse, David, "Space Rock on Collision Course," BBC News Online, 7/24/02.

Atomic Warfare

"Nuclear Fusion Basics," www.jet.efda.org/content/fusion1.html.

"Nuclear and Radiological Weapons: What's What?" www.peace-action.org/camp/starwars/swhist.html.

Asimov, Isaac, *Asimov's Chronology of Science and Discovery* (HarperCollins, 1994).

Long, Doug, "Atomic Bomb," www.doug-long.com.

Overbye, Dennis, "Our Final Hour," *New York Times*, 5/18/03.

The Big Bang

Britt, Robert, "New Theory Addresses How the Sun Was Born," Space.com, 5/20/04.

Mottram, Linda, "Big Bang Theory Challenged," Australian Broadcasting Corporation, 4/26/02.

Reich, Eugenie, "It's Just a Matter of Time," *NewScientist*, 8/21/04, pp. 34–35.

Veneziano, Gabriele, "The Myth of the Beginning of Time," *Scientific American*, 5/04.

Big Brother

"Big Brother's Following You," *NewScientist*, 2/5/05, p. 25.

"Emotion-Detecting Software," *The Week*, 2/4/04, p. 19.

"Fingerprint Systems Work" and "RFID Chips Implanted," *InformationWeek*, 7/19/04, p. 15.

"The Government is Watching You," *Bottom Line/Personal*, 8/1/04, p. 9.

"Stealth Wallpaper Keeps Company Secrets Safe," *NewScientist*, 8/7/04, p. 19.

"TV Habits Reveal Who You Are," *NewScientist*, 4/2/05, p. 18.

"We Don't Need to See Your ID," *NewScientist*, 10/23/04, p. 4.

Austen, Ian, "A Scanner Skips the ID Card and Zeroes in on the Eyes," *New York Times*, 5/15/03.

Bamford, James, "Big Brother is Tracking You Without a Warrant," *New York Times*, 5/18/03.

Biever, Celeste, "You Have Three Happy Messages," *NewScientist*, 1/8/05, p. 21.

Lewis, Holden, "Banks Are Selling Your Private Information," Bankrate.com, 10/8/02.

Monson, Suzanne, "Computer Forensics Specialists in Demand as Hacking Grows," *Seattle Times*, 9/8/02.

Phillips, Helen, "Private Thoughts, Public Property," *NewScientist*, 7/31/04, pp. 38–41.

Bionics

"Aligning Carbon Nanotubes to Aid Artificial Joints," *Chemical Engineering Progress*, 1/05, p. 15.
"Cellular Chess and the Why of Splash," *BusinessWeek*, 4/11/05, p. 95.
"Polymer Injection Helps Mend Damaged Nerves," *NewScientist*, 11/12/04, p. 9.
"Science's Potential to Create New Markets," *American Demographics*, 8/02, p. 52.
"Silicon Stitching," *Scientific American*, 1/03, p. 20.
"Silky Knee," *Technology Review*, 4/04, p. 18.
"Where's the Remote?" *Words from Woody*, Winter 2005, p. 8.
Davis, Lisa, "Fake Blood...For Real," *Reader's Digest*, 10/02, p. 77.
Donn, Jeff, "New Artificial Limbs Move Like the Real Thing," Associated Press, 6/10/01.
Fischetti, Mark, "Cochlear Implants: To Hear Again," *Scientific American*, 6/03, p. 82.
Gupta, Sanjay, "Bionics: It's Not Science Fiction Any More," CNN, 2/18/02.
Maugh, Thomas, "Sight Restoration," *The Record*, 10/14/02, p. F-1.
Marsa, Linda, "Bionic Nerve Retrains Atrophied Muscle," *The Record*, 1/20/03, p. F-3.
Nowak, Rachel, "A Better Life with a Mechanical Heart," *NewScientist*, 10/30/04, p. 28.
Ortega, Ralph, "A Miracle for Deaf 2-Year-Old," *Daily News*, 3/5/03, p. 15.
SoRelle, Ruth, "Controversies From the Heart," *Houston Chronicle*, 10/11/97.
Yonks, Jamie, "Heart Pioneer Speaks on Research," *Cornell Daily Sun*, 4/15/02.

Black Holes

"Black Hole Mystery Mimicked by Supercomputer," press releases, Jet Propulsion Laboratory, 1/24/02.
"Black Hole Sings the Deepest B-Flat," Reuters Limited, 2003.
Britt, Robert Roy, "Scientists Watch Black Hole Rip Star Apart," Space.com, 2/18/04.
Gehrels, Neil; Piro, Luigi; and Leonard, Peter, "The Brightest Explosions in the Universe," *Scientific American*, 12/02, pp. 85–91.
Stenger, Richard, "Black Hole Outburst Looks Faster Than Light," CNN, 10/3/02.
Wright, Karen, "Black Holes Made Here," *Discover*, 6/04, pp. 62–63.

Cloning

"China Boasts of Human Cloning," EWTN News, 3/7/02.
"China to Try Cloning of Rare Monkeys," Agence France-Presse, 11/25/97.
"China Successfully Clones Goats," People's Daily Online, 1/25/00.

"Cloning the Dead," *NewScientist,* 9/4/04, p. 4.

"A Mammoth Undertaking," Associated Press, 7/24/03.

"Cloning Pioneer Dolly Put to Death," MSNBC News/Associated Press, 2/14/02.

"Ma's Eyes, Not Her Ways," *Scientific American,* 4/03, p. 30.

"The Problem with Clones," *NewScientist,* 11/6/04, p. 20.

Beardsley, Tom, "A Clone in Sheep's Clothing," *Scientific American,* 4/9/02.

Crenson, Matt, "Confirmation Would Only Begin the Arguments," *Record,* 12/29/02.

Dayuan, Chen, "Clones in China," *China Today,* undated.

DeSalle, Rob and Lindley, David, *The Science of Jurassic Park and the Lost World* (Harper Perennial, 1997).

Greene, Richard, "Cloning and Genetic Engineering," *Chemical Engineering Progress,* 12/02, p. 13.

Gittings, John, "Experts Call for Curbs on Human Cloning in China," *The Guardian,* 4/16/02.

Holmes, Bob, "Ancient Genes Rise from the Dead," *NewScientist,* 12/4/04, p. 6.

———, "Squeeze Gently to Clone Monkeys," *NewScientist,* 12/11/04, p. 8.

Martindale, Diane, "Mickey Has Two Moms," *Scientific American,* 7/04, p. 24.

Mitchell, Steve, "Second Cloned Endangered Animal on the Way," UPI, 8/23/02.

Ridley, Matt, "Will We Clone a Dinosaur?" Time.com, 4/3/00.

Shermer, Michael, "I, Clone," *Scientific American,* 4/03, p. 38.

Shin, Paul, "Clone Rangers' Mule Train," *Daily News,* 5/30/02, p. 6.

Terzian, Philip, "Cloning Needs Philosopher Named Max," *Montana Standard,* 11/30/01.

———, "Send in the Clones," *Jewish World Review,* 11/29/02.

Vangelova, Luba, "True or False? Extinction is Forever," *Smithsonian,* 6/03, pp. 22–24.

Colonies in Space

Brain, Marshall and Bonsor, Kevin, "Asteroids Could Supply Moon, Mars Bases," HowStuffWorks.com, 11/10/00.

Clute, John and Nicholls, Peter, *The Encyclopedia of Science Fiction* (St. Martin's Griffin, 1995).

David, Leonard, "The Moon or Mars...Which Shall It Be," Space.com, 1/28/02.

Heppenheimer, T. A., *Colonies in Space* (Warner Books, 1977).

Communications Satellites

Communications (Time-Life Books, 1986).

Clarke, Arthur C., *Greetings, Carbon-Based Bipeds* (St. Martin's Press, 1999).

Glover, Daniel R., "NASA Experimental Communications Satellites," http://roland.nerc.nasa.gov

Whalen, David, "Communications Satellites: Making the Global Village Possible," NASA. http://www.hq.nasa.gov/office/pao/History/satcomhistory.html

Communicators

Baig, Edward, "That Enhanced Device in Your Hand Really Isn't Just a Cellphone Anymore," USA Today, 11/18/02, p. 5E.

Biersdorfer, J. D., "Hollywood's Gadget Factories," New York Times, 9//26/03, p. G1, G7.

Maney, Kevin, "Sidekick Delivers Hip-Hop Design Twist," USA Today, 11/18/02, p. 5E.

McKay, Martha, "The Latest in Wireless Gizmos," The Record, October 20, 2002, p. B-1.

Martinez, Michael, "Tomorrow's Tech Today," Kiplinger's Personal Finance, 4/02, p. 121.

Computers

Bly, Robert, Computers: Pascal, Pong & Pac-Man: A Child's History of Computers (Banbury Books, 1984).

Kroeker, Kirk and Vos Post, Jonathan, "Writing the Future: Computers in Science Fiction," IEEE Computer, 1/00, pp. 29–37.

McCartney, Scott, ENIAC (Walker and Company, 1999).

Port, Otis, "Holy Screaming Teraflops," BusinessWeek, 1/17/05, p. 62.

Cryogenic Preservation

Cryonics: Reaching For Tomorrow (Alcor Life Extension Foundation, 1993).

"Ancient Bacteria Woken From Deep Alaskan Sleep," NewScientist, 3/5/05, p. 12.

"Noah's Freezer," NewScientist, 7/31/04, p. 5.

"Psychrophile," NewScientist, 7/9/05, p. 45.

Davenport, John, Animal Life at Low Temperatures (Chapman & Hall, 1992).

Reich, Paul, "Scientists Revive Microbes from Icy Antarctic Lake," The Record, 12/17/02, p. A-4.

Slemen, Thomas, Strange But True (Barnes & Noble Books, 1998).

Wilson, Elizabeth, "How Arctic Fish Avoid Freezing," Chemical & Engineering News, 2/16/04, p. 13.

Cyborgs

"Scientists Test First Human Cyborg," CNN.com, 5/22/02.

Eisenberg, Anne, "Wired to the Brain of a Rat, a Robot Takes on the World," New York Times, 5/15/03.

Mann, Steve and Niedzviecki, Hal, *Cyborg: Digital Destiny and Human Possibility in the Age of the Wearable Computer* (Randomhouse Doubleday, 2001).

Deep Space Exploration

"Voyager Maintenance from 7 Billion Miles Away," NASA, 4/8/02.
Dyson, Freeman, "Will We Travel to the Stars?" Time.com.
Gilks, Marc; Fleming, Paula; and Allen, Moira, "Is Science Fiction For You?", *The Writer*, 11/02, pp. 34–40.
Whitehouse, David, "Life May Swim Within Distant Moons," *BBC News*, 10/02.

Dimensions, Other

Eckert, Win Scott, "Alternate Dimensions and Universes to the Wold Newton Universe," www.pjfarmer.com.
Garisto, Robert, "Curling Up Extra Dimensions in String Theory," *Physical Review Letters*, 4/8/96.

Electric Cars

"All Gassed Up," *Scientific American*, 3/04, p. 34.
"Alternative Energy Plans Focus on Hydrogen," *Chemical Engineering Progress*, 3/03, p. 23.
"Hydrogen Supply," BOC Gases brochure.
Aamot, Gregg, "Creating Energy on the Cheap," *The Record*, 2/13/04, p. A-8.
Wald, Matthew, "Questions About a Hydrogen Economy," *Scientific American*, 5/04, pp. 68–73.

Entropy

Asimov, Isaac, *Understanding Physics* (Walker and Company, 1966).
Baez, John, "The End of the Universe," http://math.ucr.edu/home/baez/end.html.
Ferris, Timothy, "How Will the Universe End?," *Time.com: Visions of the 21st Century*.

ESP (Psionics)

Haber, Karen and Yaco, Link, *The Science of the X-Men* (BP Books, 2000).

Exoskeleton

Novak, Phil, "Hurtubise Builds New Suit," *North Bay Nugget*.
Yokohama, John, "Dress for Action with Bionic Suit," *NewScientist*, 4/9/05, p. 19.

Faster-than-Light Travel

———, "Tachyons," www.physics.gmu.edu
David, Paul, *How to Build a Time Machine* (Viking Pengiun, 2001).

First Contact

"Baked Alaska Mud Volcano Discovered in North Atlantic," http://volcano. und.nodak.edu.

"ET First Contact Within 20 Years," *NewScientist*, 7/24/04, p. 24.

"Life Among Worlds Beyond Beyond," www.spacedaily.com, 2/22/02.

"Satellites of the Outer Planets," www.lpi.usra.edu.

Aczel, Amir, *Probability 1* (Harcourt, 1998).

Bridges, Andrew, "After 7 Years, Probe Rings at Saturn's Door," *The Record*, 6/12/04, p. A-16.

Hey, Nigel, "To Catch a Comet," *Smithsonian*, 1/03.

Hoagland, Richard, *The Monuments of Mars* (Frog, 1996).

Hogan, Jenny, "A Whiff of Life on the Red Planet," *NewScientist*, 2/19/05, p. 6.

Lovgren, Stefan, "Far-Out Theory Ties SARS Origins to Comet," *National Geographic News*, 6/3/03.

Overbye, Dennis, "Similar Solar System Only 90 Light Years Away," *New York Times*, 7/4/03.

Roach, John, "Microbial Colony in U.S. Suggests Life Could Live on Mars," *National Geographic News*, 1/16/02.

Flying Cars

Allen, Mike, "When the Rubber Leaves the Road," *Popular Mechanics*, 7/05, p. 72–124.

Bonsor, Kevin, "How Flying Cars Will Work," www.howstuffworks.com.

Food Pills

Lefcowitz, Eric, "Let Them Eat Fake," *Retrofuture Today*, www.retrofuture. com.

Force Fields

Krauss, Lawrence, *The Physics of Star Trek* (HarperPerennial, 1996).

Genetic Engineering

"Brave New Mouse," *Smithsonian*, 4/03, p. 19.

"Deaf Guinea Pigs Get Hearing Back," *NewScientist*, 2/19/05, p. 15.

"Fruit Takes Walk on the Wild Side," *NewScientist*, 8/28/04, p. 15.

"Salmon vs. Salmon," *Scientific American*, 8/04, p. 28.

"Scientists Deciphering Genetic Code of Microbe," People's Daily Online, 1/25/00.

"Silkworms Spin Collagen in Cocoons," *Chemical Engineering Progress*, 2/03, p. 14.

"The Trout With Salmon Fathers," *NewScientist*, 8/7/04, p. 11.

"Trials of Rice," *NewScientist*, 4/2/05, p. 7.

"With Worm Experiments, Scientists Turn Off Genes," *New York Times*, 1/15/03.

"Gene Makes Marathon Mice," *Daily News*, 8/24/04.

Aczel, Amir, *Probability 1* (Harcourt, 1998).

Cohen, Philip, "Marathon Mice Can Run and Run," *NewScientist*, 8/28/04, p. 12.

Gresh, Lois and Weinberg, Robert, *The Science of Superheroes* (John Wiley & Sons, 2002).

Haney, Daniel, "Fruit Fly Genome Decoded," ABCNEWS.com, 2/18/00.

Homeyer, Henry, "A Frog Lends a Hand to Rhododendrons," *New York Times*, 3/9/03.

Recer, Paul, "Gene Therapy OK'd for 12 with Parkinson's Disease," *The Record*, 10/11/02, p. A-10.

Smith, Ian, "New Cancer Gene," *Daily News*, 10/14/02, p. 49.

Genetically Altered Food

"Genetically Altered Foods Raise Safety Question," MSNBC.com, 6/25/02.

"Keeping Our Soldiers Fueled and Happy," *Chemical Engineering Progress*, 5/04, p. 64.

"The Secrets of Life," *Time*, 2/17/03, p. 45.

"Yes, We Have Old Bananas," *NJ Biz*, 9/8/03, p. 4.

Batalion, Nathan, "50 Harmful Effects of Genetically Modified Foods," Americans for Safe Food, www.cqs.com.

Beasley, Deena, "U.S. Opposed Labeling Genetically Engineered Food," Reuters, 6/11/02.

Chin, Kristine, *Chemical Engineering Progress*, 12/02, p. 7.

D'Aquino, Rita, *Chemical Engineering Progress*, 10/02, p. 10–15.

Giants

"In Search of Giants," www.offthefence.com

"Man Claims to be the World's Tallest," CBBC Newsround, 2/18/02.

Craddock, Bryan; Skelton, Kirk; and Wilson, Michael, "The Gentle Giant," www.mcleansboro.com.

Paterniti, Michael, "The View from Up There," *The Week*, 7/17/05.

Wolf, Buck, "High Hopes," ABCNews.com, 1/7/02.

Global Warming (the Greenhouse Effect)

"Climate Change," BBC Hot Topics, www.bbc.co.uk, 8/22/02.

"Heatwaves are Here to Stay," *NewScientist*, 8/21/04, p. 17.

"Kyoto Global Warming Pact Takes Effect," *Chemical Engineering Progress*, 3/05, p. 11.

"Phytoplankton to the Rescue," *Scientific American*, 11/02, p. 12.

"Sizzling Times Ahead for Earth," *NewScientist*, 1/29/05, p. 16.

"When the Earth Was Poisoned," *The Week*, 2/11/05, p. 21.

Appell, David, "Acting Locally," *Scientific American*, June 2003, p. 20.

Bindschadler, Robert and Bentley, Charles, "On Thin Ice," *Scientific American*, 12/02, p. 101.

Britt, Robert, "A Wild Idea to Fight Global Warming," MSNBC.com, 6/27/05.

Choi, Charles, "Hot Stuff Coming Through," *Scientific American*, 7/04, p. 36.

Eblen, Ruth and William, *The Environment Encyclopedia: Volume 1* (Marshall Cavendish, 2001).

McFarling, Usha, "Is It Just Me, or is the Planet Getting Warmer," *The Record*, 12/13/03, p. A-42.

Mitchell, Alanna, "Arctic Ice Cap Losing a Texas-Size Chunk a Decade," *The Record*, 11/30/03.

Simpson, Sarah, "Rising Sun," *Scientific American*, 6/03, p. 28.

Sturm, Matthew; Perovich, Donald; and Serreze, Mark, "Meltdown in the North," *Scientific American*, 10/03, p. 62.

Tullo, Alex, "Warming Up to Global Warming," *Chemical & Engineering News*, 2/9/04, p. 20.

Wagger, David, "Don't Avoid the Argument," *Chemical Engineering Progress*, 12/02, p. 9.

Holograms

Concise Science Dictionary (Oxford University Press, 1991).

www.holoprotec.com.

Immortality and Longevity

"Breakthrough in Premature Aging," *NewScientist*, 3/12/05, p. 16.

"Can We Prevent Aging?" *NewScientist*, 9/4/04, p. 32.

"Death Knells for Immortality," *NewScientist*, 2/65/05, p. 19.

"The Elixir of Life," *The Week*, 3/11/05, p. 21.

Borek, Carmia, "Telomere Control and Cellular Aging," *Life Extension*, 10/02, 56–59.

Duenwald, Mary, "The Puzzle of the Century," *Smithsonian*, 1/03, pp. 73–80.

Haas, Jane, "Agelessness is on the Horizon," *The Record*, 1/19/03, F-3.

Klein, Bruce, "This Wonderful Lengthening Lifespan," Longevity Meme, www.imminst.org.

Klerkx, Greg, "The Immortal's Club," *NewScientist*, 4/9/05, p. 38–41.

Kushner, Harold, *Living a Life That Matters* (Alfred A. Knopf, 2000).

Lane, David, "Dark Angel," *NewScientist*, 12/18/04, pp. 38–41.

Pearl, Raymond, "The Biology of Death: Conditions of Cellular Immortality," *The Scientific Monthly*, 4/21, p. 334.

Schermer, Michael, "Mustangs, Monsters, and Meaning," *Scientific American*, 9/04, p. 38.

Internet, the

"A Brief History of the Internet and Related Networks," www.isoc.org.
Leiner, Barry, et. al., "A Brief History of the Internet," 8/4/00, www.isoc.org.
Newton, Harry, *Newton's Telecom Dictionary* (Flatiron Publishing, 1994).
Sterling, Bruce, "Short History of the Internet," *Magazine of Fantasy & Science Fiction,* 2/93.

Jet Packs

"The Jet Flying Belt: A New Dimension in Individual Mobility," 1970, Bell Aerospace Company.
www.howstuffworks.com.
Goodwin, Harold, *All About Rockets and Space Flight* (All About Books, 1964).

Lasers and Ray Guns

"Ray Guns, Lasers in Development as Nonlethal Weapons," *St. Petersburg Times,* 11/23/02.
"Strategic Defense Initiative," www.fas.org.
Asimov, Isaac, *Asimov's Chronology of Science & Discovery* (HarperCollins, 1989).

Liquid Metal

"Movers and Shakers," *Technology Review,* 3/04, p. 16.
"Smart Fluids Solidify Market Presence," www.Thomasregional.com Industrial Market Trends.
"Probable Discovery of a New, Supersolid Phase of Matter," www.science.psu. edu, News About Eberly College of Science.
Zandonella, Catherine, "A Drop of the Hard Stuff," *NewScientist,* 4/2/05, pp. 35–38.

Men on the Moon

Clarke, Arthur C., *Greetings, Carbon-Based Bipeds* (St. Martin's, 1999).
Rupley, Sebastian, "Backup Data on the Moon," *PC Magazine,* 7/22/03.
Silver, Steven, "The First Men on the Moon," SF Site, www.sfsite.com

Mind Control

"The Future of Mind Control," *The Economist,* 5/25/02, p. 11.
www.eharassment.ca/.
Keith, Jim, "Experiments into Remote Mind Control Technology," www. karenlyster.com/keith/html.
Lee, Martin, "Truth Serums and Torture," www.alternet.org, 6/11/02.

Miniaturization and Shrinking Rays

Faucher, Elizabeth, *Honey, I Shrunk The Kids* (Scholastic, 1989).

Rovin, Jeff, *Classic Science Fiction Films*, (Carol Publishing, 1993).

Monsters

"Loch Ness Monster," *The Skeptic's Dictionary*, http://skepdic.com/nessie.html.

Mannix, Daniel, *Freaks: We Who Are Not as Others* (RE/Search, 1976).

Rovin, Jeff, *Classic Science Fiction Films* (Carol Publishing, 1993).

Mutations

Berry, Adrian, *The Book of Scientific Anecdotes* (Prometheus Books, 1993).

Blaustein, Andrew and Johnson, Pieter, "Explaining Frog Deformities," *Scientific American*, 2/03, pp. 60–65.

Choi, Charles, "When Air Quality Hits Mutant," *Scientific American*, 7/04, p. 36.

Huxley, Julian, *Evolution in Action* (New American Library, 1953).

Nanotechnology

"Biomimicking Bandages," *Chemical Engineering Progress*, 4/03, p. 10.

"Building Better Bones Through Nanotechnology," *Chemical Engineering Progress*, 11/03, p. 17.

"Carbon Nanotube Conveyor," *Chemical Engineering Progress*, 7/04, p. 16–11.

"Creating Nanostructures via Genetic Engineering," *Chemical Engineering Progress*, 4/03, pp. 16–17.

"Moving Closer to Nanotube-Based Solar Cells," *Chemical Engineering Progress*, 11/03, p. 17.

"Nano-Drug May Starve Tumors," *NewScientist*, 11/27/04, p. 17.

Nanotechnology Report, 6/02, p. 6.

"Nucleotide Nanotubes," *Scientific American*, 12/02, p. 36.

"Selective Coatings Create Biosensors from Nanotubes" and "Nanoshells Enhance Sensitivity of Chemical Detection," *Chemical Engineering Progress*, 2/05, pp. 11–12.

"Using Living Things to Build Nanomaterials," *Chemical Engineering Progress*, 8/03, p. 15.

Crichton, Michael, "Could Tiny Machines Rule the World?" *Parade Magazine*, 11/24/02, pp. 6–8.

Davies, Katharine, "Molecular Piston Shuttles Into Life," *NewScientist*, 11/20/04, p. 24.

Feynman, Richard, "There's Plenty of Room at the Bottom," *Engineering and Science*, 2/60.

Fountain, Henry, "Dishing in the Laboratory," *New York Times*, 4/29/03.

Kroeker, Kirk and Vos Post, Jonathan, "Writing the Future: Computers in Science Fiction," *IEEE Computer*, 1/00, pp. 29–37.

Ricadela, Aaron, "HP Labs Breaches a Nanotech Barrier," *InformationWeek*, 2/7/05, p. 26.

Rittner, Mindy, "Nanoparticles: What's Now, What's Next," *Chemical Engineering Progress*, 11/03, p. 39S.

Stix, Gary, "Nano Patterning," *Scientific American*, 4/04, p. 44.

Vettiger, Peter and Binnig, Gerd, "The Nanodrive Project," *Scientific American*, 1/03, pp. 47–54.

Wolfe, Josh, "Nanotechnology," *Forbes*, 3/02.

Neutron Stars

Miller, M. Coleman, "Introduction to Neutron Stars," University of Maryland. www.astro.umd.edu.

Nuclear Energy

http://hyperphysics.phy-astr.gsu.edu/hbase/nucene/fusion.html.

http://www.pppl.gov/fusion_basics/pages/fusion_power_plant.html.

Battersby, Stephen, "Fire Down Below," *NewScientist*, 8/17/04, pp. 26–29.

Muir, Hazel, "Water Regulated Natural Nuke," *NewScientist*, 11/6/04.

Nuclear War

http://zebu.uoregon.edu.

Babst, Dean, "Preventing an Accidental Nuclear War," www.wagingpeace.org.

Parallel Universes and Parallel Worlds

www.sciencenet.org.

Tegmark, Max, "Parallel Universes," *Scientific American*, 5/03, pp. 41–51.

Perpetual Motion Machines

www.zpower.net.

Lemonick, Michael, "Will Someone Build a Perpetual Motion Machine," Time.com, Visions of the 21st Century.

Milton, Richard, "Perpetual Motion," www.AlternativeScience.com

Plague, Humankind Wiped out by

"Dicing with Death," *NewScientist*, 11/20/04, p. 3.

Associated Press, "A Promising Twist on AIDS Cure," *Wired News*, 8/19/02.

Choi, Charles, "Nipah's Return," *Scientific American*, 9/04, pp. 21A-22.

Davidson, Keay, "Taking Stock of Smallpox Viruses," *San Francisco Chronicle*, 4/8/02.

Lewis, Rick, "The Rise of Antibiotic-Resistant Drugs," www.fda.gov.

Penn, Clara, "Pandemic Bug May Make a Comeback," *NewScientist*, 4/9/05, p. 8.

Robots

"We're Building a Dream One Robot at a Time," Honda ad.
McDermott, Michael, "Robot Power," *Continental*, 9/02, pp. 47–49.
Moravec, Hans, *Robot* (Oxford University Press, 1999).
Rovin, Jeff, *Classic Science Fiction Films* (Carol Publishing, 1993).
Soat, John, "U2, UN, and Robots Invade the Homeland," *InformationWeek*, 10/25/04, p. 94.
Wood, Gaby, *Edison's Eve* (Alfred A. Knopf, 2002).

Robotic Surgery

Schapiro, Robert, "Hard Wired Nurse Helps Docs Operate," *Daily News*, 6/17/05.

Rocket Ships

"Cybernetics in Industry," www.morph.demon.co.uk/Electronics/robots.htm.
Asimov, Isaac, *Asimov's Biographical Encyclopedia of Science & Technology* (Doubleday, 1964).
Clark, John, *Philip's Science & Technology: People, Dates, & Events* (George Philip Unlimited, 1999).
Clarkson, Mark, "Battlebots: The Official Behind-the-Scenes Guide," http://shop.osborne.com.
Goodwin, Harold, *All About Rockets and Space Flight* (All About Books, 1964).

Smart Weapons

"Smart Guns," New Democrats Online, www.ndol.org.
Goldstein, Scott, "Aiming for a Smart Gun," *NJ Biz*, 1/20/03, p. 6.

Space Elevator

"The Next Best Thing to a Stairway to the Stars," *BusinessWeek*, 4/18/05, p. 89.
Rabkin, Eric, "Novel Inspiration," *BusinessWeek*, 10/11/04, p. 206.

Subsurface Life

The Week, 2/25/05, p. 21.
"Microbes: Science Gets Hot Under the Crust," www.physicsweb.org.
"Project Mohole," www.nas.edu.
Fox, Douglas, "Subterranean Bugs Reach Out for Their Energy," *NewScienctist*, 6/25/05, p. 21.
Monastersky, Richard, "Deep Dwellers," *Science News*, 3/29/97.

Superbeings

"Genetic Mutation Turns Tot into Superboy," The Associated Press, 6/24/04, www.msnbc.com.

Superconductivity

"An Environmentally Friendly Route to Superconductor Production," *Chemical Engineering Progress*, 3/04, p. 13.

"Researchers Make Progress in Understanding the Basics of High-Temperature Superconductivity," http://pr.caltech/edu.

Jamieson, Valerie, "New Frontiers in Superconductivity," 1/02, physicsweb.org.

Leeb, Stephen, "Vulture Investing," *Personal Finance*, 11/13/02, p. 2.

Minkel, J. R., "Outer Quantum Limits," *Scientific American*, 6/04, p. 36.

Simmons, John, *The Scientific 100* (Carol Publishing, 1996).

Superintelligent Animals

"A Conversation with Koko," www.pbs.org.

Blackstock, Regina, "Dolphins and Man Equals," www.polaris.net.

Hooper, Rowan, "Dolphins Go to Sponge School," *NewScientist*, 6/11/05, p. 12.

Zorich, Zach, "The Gene that Made Us Human," *Discover*, 3/4/04.

Suspended Animation

Choi, Charles, "Holding in Suspense," *Scientific American*, 1/04, p. 30.

Rosenthal, Elisabeth, "Suspended Animation: Surgery's Frontier," *New York Times*, 11/13/90.

Roth, Mark, and Nystul, Todd, "Buying Time in Suspended Animation," *Scientific American*, 6/05, pp. 24–31.

Whitehouse, David, "Fish in Suspended Animation," *BBC News*, 11/23/03.

Teleportation and Transporters

Nielsen, Michael, "Rules for A Complex Quantum World," *Scientific American*, 11/02, p. 72.

Schachtman, Noah, "It's Teleportation—for Real," Wired News, www.wired.com.

Yaco, Link and Haber, Karen, *The Science of the X-Men* (BP Books, 2000).

Television

Clark, John, *Philip's Science & Technology: People, Dates, & Events* (George Philip Unlimited, 1999).

Dekhtyar, Lyudmila, "Biography of Philo T. Farnsworth," www.slcc.edu.

Terraforming

"Good Week For Desperation," *The Week*, 4/1/05, p. 8.

"Making Air on the Moon," *The Week*, 6/10/05, p. 26.

Test-Tube Babies

"U.S. Clinics Hold 400,000 Embryos," *The Record*, 5/8/03.
Henig, Robin, "Pandora's Baby," *Scientific American*, 6/03, pp. 63–67.

Time Travel

Davies, Paul, *How to Build a Time Machine* (Viking Penguin, 2001).
Gilks, Marg, "Is Science Fiction for You," *The Writer*, 9/02, p. 40.
Weiss, P., "Realistic Time Machine," ScienceNews, 7/16/05, p. 38.

Transmutation of Metals

Lockemann, Georg, *The Story of Chemistry* (Philosophical Library, 1959).

Undersea Cities

www.atlan.org/sci/.
Gresh, Lois and Weinberg, Robert, *The Science of Superheroes* (John Wiley & Sons, 2002).
Lyne, Jack, "$550 Million Underwater Hotel Launched in Dubai," *Online Insider*, www.conway.com.
Schaer, Sidney, "Suiting Up for Life Underwater," *Newsday*, undated.

Virtual Reality

Davin, Eric Leif, *Pioneers of Wonder* (Prometheus Books, 1999).
Hogan, Jenny, "Sony Patent Takes First Step to Real-Life Matrix," *NewScientist*, 4/9/05, p. 10.
Kurzweil, Ray, *The Age of Spiritual Machines* (Penguin, 2000).
Seven, Richard, "At the HIT Lab, There's Virtue in Virtual Reality," *Seattle Times Magazine*, 4/11/04.
Yeffeth, Glenn (ed.), *Taking the Red Pill: Science, Philosophy and Religion in The Matrix* (Benbella, 2003).

Water, Global Shortage of

"Earth Dries Up as Temperature Rises," *NewScientist*, 1/23/05, p. 17.
Ettouney, Hisham, "Evaluating the Economics of Desalination," *Chemical Engineering Progress*, 12/02, pp. 32–39.
Pearce, Fred, "Asian Farmers Suck the Continent Dry," *NewScientist*, 8/28/04, p. 6.
Tesar, Jenny, *Food and Water: Threats, Shortages, and Solution* (Facts on File, 1992).

Weather Control

"Electronic Weather," www.rense.com.
"The Physical Basis for Seeding Clouds," www.atmos-inc.com.
Hoffman, Ross, "Controlling Hurricanes," *Scientific American*, 10/04, p. 74.

Wormholes

"Wormholes," www.crystalinks.com/wormholes.html.
Clute, John and Nicholls, Peter, *The Encyclopedia of Science Fiction* (St. Martin's, 1995).

X-Ray Vision

Asimov, Isaac, *Words from Science* (New American Library, 1959).
Hirsh, Lou, "New Technology Allows X-ray Vision," Newsfactor Network, www.wirelessnewsfactor.com.
Tresca, Amber, "X-Rays," htto://ibscrohns.about.com.

About the Author

BOB BLY is the author of more than sixty books including *The Ultimate Unauthorized Star Trek Quiz Book* (HarperCollins), *Why You Should Never Beam Down in a Red Shirt* (HarperCollins), and *Comic Book Heroes: 1,001 Trivia Questions About America's Favorite Superheroes* (Carol Publishing Group). A science fiction fan since age twelve, he has read more than five hundred science fiction novels and stories, and seen dozens of science fiction films. Bob has sold short fiction to *Galaxy* science fiction magazine.

Bob's science credentials include a B.S. in chemical engineering and articles in such publications as *Chemical Engineering, Chemical Engineering Progress*, and *Science Books & Films*. He has been a member of the American Institute of Chemical Engineers since 1979.

Questions, comments, and suggestions for future editions may be sent to:

Bob Bly
22 E. Quackenbush Avenue
Dumont, NJ 07628
Web: www.bly.com
E-mail: rwbly@bly.com

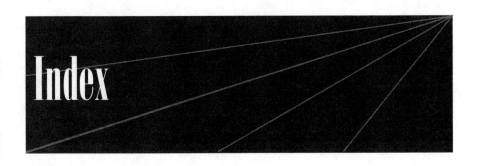

Index

X